Lecture Notes in Mathematics

Edited by A. Dold and B. Eckmann

1347

Chris Preston

Iterates of Piecewise Monotone Mappings on an Interval

Springer-Verlag

Berlin Heidelberg New York London Paris Tokyo

Author

Chris Preston
FSP Mathematisierung, Universität Bielefeld
4800 Bielefeld 1, Federal Republic of Germany

Mathematics Subject Classification (1980): 58F08, 54H20, 26A18

ISBN 3-540-50329-3 Springer-Verlag Berlin Heidelberg New York
ISBN 0-387-50329-3 Springer-Verlag New York Berlin Heidelberg

Printing and binding: Druckhaus Beltz, Hemsbach/Bergstr.
2146/3140-543210

Piecewise monotone mappings on an interval provide simple examples of discrete dynamical systems whose behaviour can be very complicated. These notes are concerned with some of the properties of such mappings. It is hoped that the material presented can be understood by anyone who has had a basic course in (one-dimensional) real analysis. This account is self-contained, but it can be regarded as a sequel to *Iterates of maps on an interval* (Springer Lecture Notes in Mathematics, Vol. 999).

I would like to thank Lai-Sang Young, Richard Hohmann-Damaschke and Jürgen Willms for their suggestions and comments during the writing of these notes. My thanks go also to the staff of the *FSP Mathematisierung* at the University of Bielefeld for technical assistance in the preparation of the text.

Bielefeld

May 1987

Chris Preston

ITERATES OF PIECEWISE MONOTONE MAPPINGS ON AN INTERVAL - CONTENTS

1. INTRODUCTION

Let $I = [a,b]$ be a closed, bounded interval and let $C(I)$ denote the set of continuous functions $f : I \to I$ which map the interval I back into itself. For $f \in C(I)$ we define $f^n \in C(I)$ inductively by $f^0(x) = x$, $f^1(x) = f(x)$ and (for $n > 1$) $f^n(x) = f(f^{n-1}(x))$. f^n is called the **n th. iterate** of f. The set of iterates $\{f^n\}_{n\geq 0}$ of a mapping $f \in C(I)$ provides us with a very simple example of a dynamical system. This system can be thought of as describing some process, whose states are represented by the points of the interval I, and which is observed at discrete time intervals (say once a year or every ten minutes); the process evolves in such a way that, if at some observation time the process is in the state x, then at the next observation time the process will be in the state $f(x)$. Hence if the process is originally in the state x (at time 0) then it will be in the state $f^n(x)$ at time n. The sequence $\{f^n(x)\}_{n\geq 0}$ is called the **orbit** of x (under f), and it describes the successive states of the process, given that x was the starting state.

Dynamical systems of this type have been used as models in the biological sciences (see, for example, May (1976), May and Oster (1976) and Guckenheimer, Oster and Ipaktchi (1977)), as well as in the physical sciences (see, for example, Lorenz (1963), Collet and Eckmann (1980) and Gumowski and Mira (1980)). They are also ideally suited for making numerical "experiments" using a computer (see, for example, Feigenbaum (1978) and (1979)).

In these notes we study the iterates of a special class of mappings in $C(I)$, namely the iterates of piecewise monotone mappings. A mapping $f \in C(I)$ is called **piecewise monotone** if there exists $N \geq 0$ and $a = d_0 < d_1 < \cdots < d_N < d_{N+1} = b$ such that f is strictly monotone on $[d_k, d_{k+1}]$ for each $k = 0, \ldots, N$. The set of piecewise monotone mappings in $C(I)$ will be denoted by $M(I)$. If $f \in M(I)$ then $w \in (a,b)$ is called a **turning point** of f if f is not monotone in any neighbourhood of w. $M(I)$ includes all of the mappings in $C(I)$ whose iterates have been used as models of "real" processes; in fact, most such models use mappings having one turning point, for example the elements in the family $p_\mu \in M([0,1])$, $0 < \mu \leq 4$, where $p_\mu(x) = \mu x(1-x)$.

We will analyse the iterates $\{f^n\}_{n\geq 0}$ for a general mapping $f \in M(I)$. The analysis is carried out in two stages. In Sections 2 to 6 we will be concerned with following question for an element f from $M(I)$: What does the asymptotic behaviour of the orbit $\{f^n(x)\}_{n\geq 0}$ look like for a "typical" point $x \in I$? Here "typical" is meant in a topological (rather than in a measure-theoretical) sense: we want to make statements about the asymptotic behaviour of $\{f^n(x)\}_{n\geq 0}$ which hold for all points x lying in a "large" subset of I , where a "large" subset is one which contains a dense open subset of I , or at least is a residual subset of I . (A residual subset is one which contains a countable intersection of dense open subsets of I .) The main result of the first stage of the analysis is Theorem 2.4; this is a generalization of Theorem 5.2 in Preston (1983) , which dealt with the case of mappings having only one turning point and which was based on results of Guckenheimer (1979) and Misiurewicz (1981). Theorem 2.4 says roughly that if $f \in M(I)$ then one of three things happens to the orbit $\{f^n(x)\}_{n\geq 0}$ of a "typical" point $x \in I$:

(1.1) The orbit eventually ends up in an f-invariant subset $C \subset I$, C consisting of finitely many closed intervals, on which f acts topologically transitively (which means that the orbit of some point in C is dense in C).

(1.2) The orbit is attracted to an f-invariant Cantor-like set $R \subset I$, on which f acts minimally (which means that the orbit of each point in R is dense in R).

(1.3) The orbit is contained in an f-invariant open set $Z \subset I$, which is such that on each of its connected components f^n is monotone for each $n \geq 0$.

(A subset $A \subset I$ is called **f-invariant** if $f(A) \subset A$.)

The second stage of the analysis is carried out in Sections 7 to 12. This can be seen as a study of the structure of the set of points $x \in I$ for which none of (1.1), (1.2) and (1.3) hold. By Theorem 2.4 this set is "small"; however, it turns out that the global complexity of the iterates of f can be strongly influenced by the behaviour of f on this set.

The aim of these notes is to analyse the topological structure of the iterates of a mapping $f \in M(I)$. This means that we are only interested in results which are invariant under topological equivalence. To make this more precise we need a definition: Mappings $f, g \in \mathbf{M}(I)$ are said to be **conjugate** if there exists a homeomorphism $\psi : I \to I$ (i.e. ψ is a continuous and strictly monotone mapping of I onto itself) such that $\psi \circ f = g \circ \psi$. The results which we present are all invariant under conjugacy; i.e. if $f, g \in \mathbf{M}(I)$ are conjugate then any statement which occurs in a theorem is either satisfied by both of f and g or it is satisfied by neither of them.

The topological structure of the iterates of a mapping $f \in \mathbf{M}(I)$ is also analysed in Milnor and Thurston (1977) and in Nitecki (1982). The present notes have little in common with the first of these papers, except for Section 6, the main part of which is only a slight modification of material in Milnor and Thurston. Nitecki not only considers mappings from $\mathbf{M}(I)$, but also the iterates of a general mapping from $\mathbf{C}(I)$. Section 4 of Nitecki's paper deals with a "spectral decomposition" theorem for the non-wandering set of a mapping $f \in \mathbf{M}(I)$, a result which is due to Hofbauer (1981) (and for mappings with only one turning point to Jonker and Rand (1981) and van Strien (1981)). The second stage of our analysis is likewise concerned with a kind of "spectral decomposition" (though not for the non-wandering set), and in a certain sense our approach parallels that taken by Nitecki. We strongly recommend the reader to study Nitecki's paper, and, if it can be got hold of, that of Milnor and Thurston. The book by Collet and Eckmann (Collet and Eckmann (1980)), which mainly treats mappings having a single turning point, is also highly recommended.

Some of the results in these notes can be extended to mappings which are allowed discontinuities at their turning points, (i.e. to the set of mappings $f : I \to I$ for which there exists $N \geq 0$ and $a = d_0 < d_1 < \cdots < d_N < d_{N+1} = b$ such that f is strictly monotone and continuous on each of the open intervals (d_k, d_{k+1}) , $k = 0, \dots, N$). Results of this type corresponding to the first stage of the analysis given here can be found in Willms (1987). Furthermore, Hofbauer's "spectral decomposition" theorem (Hofbauer (1981), (1986)) remains valid in this extended set-up.

As we have already mentioned, we hope that the material in these notes can understood by anyone who has had a basic course in real analysis. The most important prerequisite is a familiarity with the standard topological properties of the real line (as are covered, for example, in the first five chapters of Rudin (1964)). There are, however, a couple of results which we need to use, and which probably do not occur in a typical introductory real analysis course (for example, the Baire category theorem). These results are stated and proved in Section 13.

Throughout these notes I denotes the closed bounded interval $[a,b]$, on which most of the mappings are defined; the symbols a and b are only ever used for the end-points of this interval. We consider I as our basic topological space; hence "open" always means with respect to the topology on I, and so, for example, $[a,c)$ is an open subset of I for each $c \in (a,b]$. If $A \subset I$ then $\bar{A}$ denotes the closure of A and $\text{int}(A)$ the interior of A (again with respect to the topology on I). If $J \subset I$ is an interval then $|J|$ will denote the length of J. (We also use $|A|$ to denote the cardinality of the set A; however, this should not create any problems.)

We now give an outline of what is contained in the various sections of these notes.

Section 2: Piecewise monotone mappings This section introduces the basic definitions and facts about piecewise monotone mappings. For $f \in M(I)$ let $T(f)$ denote the set of turning points of f and let

$$Z(f) = \{ x \in (a,b) : \text{there exists } \varepsilon > 0 \text{ such that } f^n \text{ is monotone on } (x-\varepsilon, x+\varepsilon) \text{ for all } n \geq 0 \} ;$$

then $Z(f)$ is open, and in fact $Z(f)$ is the largest open set $G \subset (a,b)$ such that $f^n(G) \cap T(f) = \varnothing$ for all $n \geq 0$. Moreover, it is not hard to see that $Z(f)$ is f-invariant. ($Z(f)$ is the set which occurs in (1.3).)

Let $m \geq 1$; we call a closed set $C \subset I$ an **f-cycle** with period m if C is the disjoint union of non-trivial closed intervals $B_0, \ldots, B_{m-1}$ such that $f(B_{k-1}) \subset B_k$ for $k = 1, \ldots, m-1$ and $f(B_{m-1}) \subset B_0$; (in particular we then have $f(C) \subset C$). We call C **proper** if $f(B_{k-1}) = B_k$ for $k = 1, \ldots, m-1$ and $f(B_{m-1}) = B_0$. An f-cycle C is said to be **topologically transitive** if whenever F is a closed subset of C with $f(F) \subset F$ then either $F = C$ or $\text{int}(F) = \varnothing$. There are several other definitions of "topologically transitive" which are equivalent to the one we have used, and some of these are given in Proposition 2.8; for example, an f-cycle C is topologically transitive if and only if the orbit of x is dense in C for some $x \in C$. (The subset C which occurs in (1.1) is a topologically transitive f-cycle.)

Let $\{C_n\}_{n\ge1}$ be a decreasing sequence of f-cycles, C_n having period m_n; it is then easy to see that $m_n | m_{n+1}$ for each $n \ge 1$. We call the sequence $\{C_n\}_{n\ge1}$ **splitting** if $m_{n+1} > m_n$ for each $n \ge 1$. We say that $R \subset I$ is an **f-register-shift** if there exists a splitting sequence of proper f-cycles $\{K_n\}_{n\ge1}$ with $K_1 \cap Z(f) = \emptyset$ such that $R = \bigcap_{n\ge1} K_n$, and we then call $\{K_n\}_{n\ge1}$ a **generator** for R. If R is an f-register-shift then Propositions 2.2 and 2.10 show that $\mathrm{int}(R) = \emptyset$ and that f maps R homeomorphically back onto itself; moreover, the orbit of x is dense in R for each $x \in R$. (The subset R which occurs in (1.2) is an f-register-shift.)

If C is an f-cycle then let

$$A(C,f) = \{ x \in I : f^n(x) \in \mathrm{int}(C) \text{ for some } n \ge 0 \}.$$

If R is an f-register-shift and $\{K_n\}_{n\ge1}$, $\{K'_n\}_{n\ge1}$ are two generators for R then it follows from Proposition 2.2 that $\bigcap_{n\ge1} A(K_n,f) = \bigcap_{n\ge1} A(K'_n,f)$; we can thus define $A(R,f) = \bigcap_{n\ge1} A(K_n,f)$. (In Proposition 2.9 we will see that

$$A(R,f) = \{ x \in I : \lim_{n\to\infty} \min_{z\in R} |f^n(x)-z| = 0 \}.)$$

Now let C, C' be topologically transitive f-cycles and R, R' be f-register-shifts. Proposition 2.2 will show that either $C = C'$ or $A(C,f) \cap A(C',f) = \emptyset$, and that either $R = R'$ or $A(R,f) \cap A(R',f) = \emptyset$; also $A(C,f)$, $A(R,f)$ and $Z(f)$ are all disjoint. Furthermore, each of $A(C,f)$ and $A(R,f)$ contains a turning point of f. Therefore if $C_1,\dots,C_r$ are the topologically transitive f-cycles and $R_1,\dots,R_\ell$ are the f-register-shifts, then $\ell+r \le |T(f)|$ and the sets $A(C_1,f),\dots,A(C_r,f)$, $A(R_1,f),\dots,A(R_\ell,f)$ and $Z(f)$ are disjoint. The first main result of Section 2 is Theorem 2.4, which says that

$$\Lambda(f) = A(C_1,f) \cup \cdots \cup A(C_r,f) \cup A(R_1,f) \cup \cdots \cup A(R_\ell,f) \cup Z(f)$$

is a dense subset of I, (and thus $\Lambda(f)$ is residual, since it can clearly be written as a countable intersection of open subsets of I). Theorem 2.4 is the precise formulation of the statement that for the orbit of a "typical" point in I one of (1.1), (1.2) and (1.3) holds.

The second main result of Section 2 (Theorem 2.5) gives us more information about the topologically transitive f-cycles. We say that $f \in \mathbf{M}(I)$ is **(topologically) exact** if for each non-trivial interval $J \subset I$ there exists $n \geq 0$ such that $f^n(J) = I$. We call $f \in \mathbf{M}(I)$ **semi-exact** if there exists $c \in (a,b)$ such that $f([a,c]) = [c,b]$, $f([c,b]) = [a,c]$, and the restriction of f^2 to $[a,c]$ is exact. Let $f \in \mathbf{M}(I)$ and C be a proper f-cycle with period m ; let B be one of the m components of C . We say that C is **exact** (resp. **semi-exact**) if the restriction of f^m to B is exact (resp. semi-exact). (It is easy to see that these definitions do not depend on which of the components of C is used.) If C is either exact or semi-exact then C is clearly topologically transitive. Theorem 2.5 states that the converse of this is true, namely that each topologically transitive f-cycle is either exact or semi-exact.

Let $f \in \mathbf{M}(I)$ be topologically transitive, (by which we mean that the whole interval I is a topologically transitive f-cycle). Then Theorem 2.5 gives us that f is either exact or semi-exact. An important property of such mappings is provided by a result of Parry (Corollary 3 in Parry (1966)). This implies that if $f \in \mathbf{M}(I)$ is either exact or semi-exact then f is conjugate to a uniformly piecewise linear mapping $g \in \mathbf{M}(I)$, where a mapping $g \in \mathbf{M}(I)$ is said to be **uniformly piecewise linear with slope** $\beta > 0$ if, on each of the intervals where it is monotone, g is linear with slope either β or $-\beta$. Thus if $f \in \mathbf{M}(I)$ is topologically transitive then f is conjugate to a uniformly piecewise linear mapping. We give a proof of this result in Section 6.

Section 3: Theorems 2.4 and 2.5 are proved in this section.

Section 4: Sinks and homtervals Here we analyse the set $Z(f)$. For $f \in \mathbf{M}(I)$ let

$$Z_*(f) = \{ x \in I : f^n(x) \in Z(f) \text{ for some } n \geq 0 \} ;$$

then $Z_*(f)$ is open, $Z(f) \subset Z_*(f)$ and $f(Z_*(f)) \subset Z_*(f)$. In fact there is not much difference between $Z(f)$ and $Z_*(f)$, since in Proposition 4.1 we show that each point in $Z_*(f)-Z(f)$ is an isolated point of $I-Z(f)$, and so in particular $Z_*(f)-Z(f)$ is countable. $Z_*(f)$ can be described in terms of sinks and homtervals. A non-empty open interval $J \subset (a,b)$ is called a **sink** of f if there exists $m \geq 1$ such that f^m is monotone on J and $f^m(J) \subset J$. (If J is a sink of f then it follows that f^n is monotone on J for all $n \geq 0$.) A non-empty open interval $J \subset (a,b)$ is called a **homterval** of f if for each $n \geq 0$ we have f^n is monotone on J and $f^n(J)$ is not contained in any sink. Let

$$\mathrm{Sink}(f) = \{ x \in I : f^n(x) \in J \text{ for some sink } J \text{ and some } n \geq 0 \},$$

$$\mathrm{Homt}(f) = \{ x \in I : f^n(x) \in L \text{ for some homterval } L \text{ and some } n \geq 0 \}.$$

$\mathrm{Sink}(f)$ and $\mathrm{Homt}(f)$ are clearly both open. We will see that both these sets are f-invariant, $\mathrm{Sink}(f) \cap \mathrm{Homt}(f) = \varnothing$ and $\mathrm{Sink}(f) \cup \mathrm{Homt}(f) = Z_*(f)$.

For $f \in M(I)$ and $n \geq 1$ let $\mathrm{Per}(n,f)$ denote the set of **periodic points** of f with **period** n, i.e.

$$\mathrm{Per}(n,f) = \{ x \in I : f^n(x) = x,\ f^k(x) \neq x \text{ for } k = 1,\dots, n-1 \}.$$

There is a strong connection between sinks and "attracting" periodic points. To see this, we need a couple of definitions. Let $f \in M(I)$; for $x \in \mathrm{Per}(m,f)$ put $[x] = \{x, f(x), \dots, f^{m-1}(x)\}$, so $[x]$ is the periodic orbit containing x. Let $\alpha([x],f)$ denote the set of points in I which are attracted to the orbit $[x]$, i.e. $\alpha([x],f) = \{ y \in I : \lim_{n\to\infty} f^{mn}(y) = f^k(x) \text{ for some } 0 \leq k < m \}$. Also let $\delta([x],f) = \{ y \in I : f^n(y) = x \text{ for some } n \geq 0 \}$; thus $\delta([x],f)$ consists of those points in I which eventually hit the orbit $[x]$. Clearly $\delta([x],f) \subset \alpha([x],f)$, and $\delta([x],f)$ is countable (because $f^{-1}(\{z\})$ is finite for each $z \in I$). Now let $f \in M(I)$ and $x \in \mathrm{Per}(m,f)$; then Proposition 4.3 shows that either

(1) $\alpha([x],f) = \delta([x],f)$, (in which case $\alpha([x],f)$ is countable), or

(2) there exists a non-trivial interval J with $x \in J \subset \alpha([x],f)$ such that $\mathrm{int}(J)$ is a sink of f.

Moreover, if *(2)* holds then $\alpha([x],f) - \delta([x],f)$ is a non-empty open subset of $\mathrm{Sink}(f)$. We say that x is **attracting** if *(2)* holds.

Let $f \in \mathbf{M}(I)$ and M be the set of attracting periodic points of f (and note that M is countable). Put $\alpha(f) = \bigcup_{x \in M} \alpha([x],f)$. Proposition 4.4 states that if $\mathrm{Per}(m,f)$ is finite for each $m \geq 1$ then $\mathrm{Sink}(f) - \alpha(f)$ and $\alpha(f) - \mathrm{Sink}(f)$ are both countable, and so in this case there is not much difference between the sets $\mathrm{Sink}(f)$ and $\alpha(f)$. The assumption in Proposition 4.4 (that $\mathrm{Per}(m,f)$ be finite for each $m \geq 1$) will clearly be satisfied if f is the restriction to I of an analytic function defined in some (complex) neighbourhood of I . Mappings which have been used in applications are usually of this type (for example, polynomial mappings).

Most of Section 4 is involved with a general analysis of the sinks of a mapping $f \in \mathbf{M}(I)$. This is very straighforward, and the material is taken, with only minor modifications, from Section 5 of Preston (1983), which in turn was based on results and ideas from Guckenheimer (1979), Collet and Eckmann (1980) and Misiurewicz (1981).

Section 5: Examples of register-shifts Here we give a couple of simple examples to demonstrate that register-shifts actually occur. The mappings we consider are not smooth, but they have the advantage of being defined explicitly, and they are very easy to analyse.

Our first example of a mapping having a register-shift is similar to one occurring in Milnor and Thurston (1977). Let $I = [0,1]$, and for each $n \geq 0$ let $a_n = \frac{1}{2}(1-3^{-n})$ and put $f(a_n) = \frac{4}{5}(1-6^{-n})$. Now define f to be linear on each interval $[a_n,a_{n+1}]$, $n \geq 0$ (and so f has slope 2^{-n+1} on $[a_n,a_{n+1}]$). This defines f as a strictly increasing, continuous function on $[0,\frac{1}{2})$. Put $f(\frac{1}{2}) = \frac{4}{5}$ and let $f(x) = f(1-x)$ for $x \in (\frac{1}{2},1]$. Then $f \in \mathbf{M}([0,1])$, $f(0) = f(1) = 0$ and $\frac{1}{2}$ is the single turning point of f .

The reason for defining f in this way is that we then have the following "self-similarity" property (which is proved in Lemma 5.2): $f^2([\frac{1}{3},\frac{2}{3}]) \subset [\frac{1}{3},\frac{2}{3}]$, and if g is the restriction of f^2 to $[\frac{1}{3},\frac{2}{3}]$ then $g(x) = \frac{1}{3}(2 - f(2-3x))$ for all $x \in [\frac{1}{3},\frac{2}{3}]$, i.e. $g = \psi \circ f \circ \psi^{-1}$, where $\psi : [0,1] \to [\frac{1}{3},\frac{2}{3}]$ is the linear change of variables given by $\psi(t) = \frac{1}{3}(2-t)$. (This means that g is just f turned upside down and scaled down by a factor of 3 .) Using this property and induction we construct a decreasing sequence of proper f-cycles $\{K_n\}_{n\geq 1}$ with $per(K_n) = 2^n$ for each $n \geq 1$. In Lemma 5.4 we show that $Z(f) = \emptyset$, and hence $R = \bigcap_{n\geq 1} K_n$ is an f-register-shift. Moreover, it follows from Lemma 5.3 that $[0,1] - A(R,f)$ is countable.

Let $f \in \mathbf{M}(I)$ and R be an f-register-shift; we say that R is **tame** if there exists an f-cycle K with $R \subset K$ such that $K - A(R,f)$ is countable. (In Section 9 we show that the structure of a tame register-shift is somewhat special.) The example given above is thus tame. However, it is easy to modify this example to obtain a non-tame register-shift. In fact, again let $I = [0,1]$, and for $n \geq 0$ let $a_n = \frac{1}{2}(1-7^{-n})$ and $f(a_n) = \frac{24}{27}(1-28^{-n})$. Define f to be linear on $[a_n,a_{n+1}]$, $n \geq 0$ (and so f has slope $2\cdot 4^{-n}$ on $[a_n,a_{n+1}]$). As before this defines f as a strictly increasing, continuous function on $[0,\frac{1}{2})$. Put $f(\frac{1}{2}) = \frac{24}{27}$ and for $x \in (\frac{1}{2},1]$ let $f(x) = f(1-x)$. This mapping f also has a "self-similarity" property, namely: $f^3([\frac{3}{7},\frac{4}{7}]) \subset [\frac{3}{7},\frac{4}{7}]$, and if g is the restriction of f^3 to $[\frac{3}{7},\frac{4}{7}]$ then $g(x) = \frac{1}{7}(4 - f(4-7x))$. (This means that g is just f turned upside down and scaled down by a factor of 7 .) As in the first example, this property allows us to construct a decreasing sequence of proper f-cycles $\{K_n\}_{n\geq 1}$, this time with $per(K_n) = 3^n$ for each $n \geq 1$. Again we have $Z(f) = \emptyset$, and so $R = \bigcap_{n\geq 1} K_n$ is an f-register-shift. However, we show that in this example R is not tame.

The reason that the register-shift in the second example is not tame is because the interval $[\frac{3}{7},\frac{4}{7}]$ sits in $[0,1]$ in a complicated way; more precisely, the set $\{ x \in [0,1] : f^n(x) \notin [\frac{3}{7},\frac{4}{7}]$ for all $n \geq 0 \}$ contains a Cantor-like set, and is thus uncountable. This is in contrast to the first example, where the corresponding set (i.e. $\{ x \in [0,1] : f^n(x) \notin [\frac{1}{3},\frac{2}{3}]$ for all $n \geq 0 \}$) consists only of the two points 0 and 1 .

Section 6: A proof of Parry's theorem In this section we give a proof of the following result in Parry (1966):

Theorem 6.1 If $f \in \mathbf{M}(I)$ is topologically transitive then f is conjugate to a uniformly piecewise linear mapping.

We do not use Parry's proof, but instead one taken, with a few minor modifications, from Milnor and Thurston (1977).

Let $V(I) = \{ \psi \in C(I) : \psi$ is increasing and onto $\}$ (where by increasing we mean only that $\psi(x) \geq \psi(y)$ whenever $x \geq y$). If $\psi \in V(I)$ and $g \in \mathbf{M}(I)$ then we say that (ψ,g) is a **reduction** (or semi-conjugacy) of $f \in \mathbf{M}(I)$ if $\psi \circ f = g \circ \psi$. Lemma 6.2 states that if $f \in \mathbf{M}(I)$ is topologically transitive and (ψ,g) is a reduction of f then ψ is automatically a homeomorphism, and so in particular f and g are conjugate. Lemma 6.2 reduces the proof of Theorem 6.1 to showing that if $f \in \mathbf{M}(I)$ is topologically transitive then there exists a reduction (ψ,g) of f with g uniformly piecewise linear.

For $f \in \mathbf{M}(I)$ let $\ell(f) = |T(f)| + 1$, and also let $h(f) = \inf_{n\geq 1} \frac{1}{n} \log \ell(f^n)$; thus $h(f) \geq 0$. Lemma 6.3 shows that in fact $h(f) = \lim_{n\to\infty} \frac{1}{n} \log \ell(f^n)$ for each $f \in \mathbf{M}(I)$, and in Lemma 6.4 we show that $h(f) > 0$ whenever f is topologically transitive. Theorem 6.1 therefore follows from Theorem 6.5, which says that if $f \in \mathbf{M}(f)$ with $h(f) > 0$ then there exists a reduction (ψ,g) of f such that g is uniformly piecewise linear with slope β , where $\beta = \exp(h(f))$.

Theorem 6.5 and its proof are due to Milnor and Thurston. The proof goes roughly as follows: Fix $f \in \mathbf{M}(I)$ with $h(f) > 0$ and put $r = \exp(-h(f))$; thus $r = 1/\beta$ and $0 < r < 1$. Note that by Lemma 6.3 we have $\beta = \lim_{n\to\infty} \ell(f^n)^{1/n}$, and hence r is the radius of convergence of the power series $\sum_{n\geq 0} \ell(f^n)t^n$; in particular the series $L(t) = \sum_{n\geq 0} \ell(f^n)t^n$ converges for all $t \in (0,r)$. Now let $J \subset I$ be a non-trivial closed interval, and for $n \geq 0$ let $\ell(f^n|J) = |T(f^n) \cap \mathrm{int}(J)| + 1$. Then $\ell(f^n|J) \leq \ell(f^n|I) = \ell(f^n)$, and therefore the series $L(J,t) = \sum_{n\geq 0} \ell(f^n|J)t^n$ also converges for all $t \in (0,r)$. Hence we can define $\Lambda(J,t) = L(J,t)/L(I,t)$ for each $t \in (0,r)$ (since $L(I,t) = L(t) \neq 0$), and we have $0 \leq \Lambda(J,t) \leq 1$, because $L(J,t) \leq L(I,t)$. In Lemma 6.10 it is shown that there exists a sequence $\{t_n\}_{n\geq 1}$ from $(0,r)$ with $\lim_{n\to\infty} t_n = r$, and such that $\{\Lambda(J,t_n)\}_{n\geq 1}$ converges for each non-trivial closed interval $J \subset I$. For each such interval J we can thus define $\Lambda(J) = \lim_{n\to\infty} \Lambda(J,t_n)$, and this gives us a mapping $\pi : I \to [0,1]$ obtained by letting $\pi(a) = 0$ and $\pi(x) = \Lambda([a,x])$ for $x \in (a,b]$.

Lemmas 6.11, 6.12 and 6.13 show that the mapping $\pi : I \to [0,1]$ is continuous, increasing and onto, that there exists a unique mapping $\alpha : [0,1] \to [0,1]$ with $\pi \circ f = \alpha \circ \pi$, and that this mapping α is uniformly piecewise linear with slope β. Thus, by a simple linear rescaling of α and π, we get a reduction (ψ, g) of f such that $g \in \mathbf{M}(I)$ is also uniformly piecewise linear with slope β.

At the end of Section 6 we give a result from Misiurewicz and Szlenk (1980), which provides an alternative method of calculating $h(f)$ for a mapping $f \in \mathbf{M}(I)$. For $f \in \mathbf{C}(I)$ let

$$\mathrm{Var}(f) = \sup \{ \sum_{k=0}^{n-1} |f(x_{k+1}) - f(x_k)| : a = a_0 < x_1 < \cdots < x_n = b \} .$$

In particular, if $f \in \mathbf{M}(I)$ has turning points $d_1, \ldots, d_N$, where $a = d_0 < d_1 < \cdots < d_N < d_{N+1} = b$, then clearly $\mathrm{Var}(f) = \sum_{k=0}^{N} |f(d_{k+1}) - f(d_k)|$. The result of Misiurewicz and Szlenk says that if $f \in \mathbf{M}(I)$ then $h(f) > 0$ if and only if $\limsup_{n\to\infty} \frac{1}{n} \log \mathrm{Var}(f^n) > 0$; moreover, $h(f) = \lim_{n\to\infty} \frac{1}{n} \log \mathrm{Var}(f^n)$ whenever $h(f) > 0$. Consider the special case of a uniformly piecewise linear mapping $g \in \mathbf{M}(I)$ with slope $\beta \geq 1$; then this result shows that $h(g) = \log \beta$, (since g^n is uniformly piecewise linear with slope β^n, and so $\mathrm{Var}(g^n) = (b-a)\beta^n$ for each $n \geq 1$).

Section 7: Reductions In this section we start the second stage in the analysis of the iterates of a mapping $f \in \mathbf{M}(I)$. Up to now the main result has been Theorem 2.4, which gives us information about the asymptotic behaviour of the orbits $\{f^n(x)\}_{n\geq 0}$ for all points x lying in a residual subset $\Lambda(f)$ of I. The set $I - \Lambda(f)$, being the complement of a residual set, is topologically "small"; however, the behaviour of f on this set can strongly influence the global complexity of the iterates of f. In order to study this we look at the reductions of f: Recall from Section 6 that if $\psi \in \mathbf{V}(I)$ and $g \in \mathbf{M}(I)$ then (ψ, g) is a reduction of f if $\psi \circ f = g \circ \psi$. Let (ψ, g) be a reduction of $f \in \mathbf{M}(I)$; then in a certain sense g describes the behaviour of f on $\mathrm{supp}(\psi)$, where

$$\mathrm{supp}(\psi) = \{ x \in I : \psi(J) \text{ is non-trivial for each open interval } J \subset I \text{ with } x \in J \} .$$

Thus the properties of g will give us information about f on $\mathrm{supp}(\psi)$. This section deals mainly with the general properties of reductions; the results obtained here will be used in the following sections.

For $f \in \mathbf{M}(I)$ put $S(f) = T(f) \cup \{a,b\}$; we say that $A \subset I$ is **f-almost-invariant** if $f(A-S(f)) \subset A$. $\mathbf{D}(f)$ will denote the set of non-empty perfect subsets D of I such that D and $I-D$ are both f-almost-invariant. ($D \subset I$ is called **perfect** if it is closed and contains no isolated points; note that if D is perfect then D is f-almost-invariant if and only if it is f-invariant.) Let $f \in \mathbf{M}(I)$ and $D \subset I$; in Theorem 7.4 we show there exists a reduction (ψ,g) of f with $\mathrm{supp}(\psi) = D$ if and only if $D \in \mathbf{D}(f)$. Moreover, the reduction (ψ,g) is essentially determined by $\mathrm{supp}(\psi)$: If (ψ',g') is another reduction of f with $\mathrm{supp}(\psi') = \mathrm{supp}(\psi)$ then there exists a unique homeomorphism $\theta \in V(I)$ such that $\psi' = \theta\circ\psi$, and $\theta\circ g = g'\circ\theta$.

It is quite easy to obtain elements of $\mathbf{D}(f)$. If A is a closed subset of I then let $\kappa(A)$ denote the set of condensation points of A , i.e. $x \in \kappa(A)$ if each neighbourhood of x contains uncountably many elements of A . Then $\kappa(A) \subset A$, $\kappa(A)$ is perfect, and $A-\kappa(A)$ is countable. (These facts constitute the Cantor-Bendixson theorem; see Lemma 7.5.) In particular, if A is uncountable then $\kappa(A)$ is non-empty. Let $f \in \mathbf{M}(I)$ and A be an uncountable closed subset of I ; we show that $\kappa(A) \in \mathbf{D}(f)$ if and only if A and $I-A$ are both f-weakly-invariant, where we define $B \subset I$ to be **f-weakly-invariant** if $f(B)-B$ is countable.

The above results can be applied with $A = I-Z(f)$, since $Z(f)$ is f-invariant and $I-Z(f)$ is f-almost-invariant. Hence if $f \in \mathbf{M}(I)$ and $I-Z(f)$ is uncountable then there exists a reduction (ψ,g) of f with $\mathrm{supp}(\psi) = \kappa(I-Z(f))$. We therefore have $\mathrm{supp}(\psi) \subset I-Z(f)$ and $(I-Z(f)) - \mathrm{supp}(\psi)$ is countable, and Proposition 7.9 will show that $Z(g) = \varnothing$. Thus (ψ,g) essentially "kills off" the set $Z(f)$; in many cases g will provide information about the iterates of f which was "obscured" by $Z(f)$. Another application is when we have $f \in \mathbf{M}(I)$ and an f-cycle C . If $I-A(C,f)$ is uncountable then there exists a reduction (ψ,g) of f with $\mathrm{supp}(\psi) = \kappa(I-A(C,f))$, because $I-A(C,f)$ is f-invariant and $A(C,f)$ is f-almost-invariant. This fact will play an important rôle in Section 8.

Section 8: The structure of the set D(f) This section continues the study of reductions by examining the structure of the set $\mathbf{D}(f)$ for a mapping $f \in \mathbf{M}_0(I)$, where $\mathbf{M}_0(I) = \{ f \in \mathbf{M}(I) : Z(f) = \emptyset \}$. The reason for making this restriction to mappings from $\mathbf{M}_0(I)$ is because it greatly simplifies the analysis, and because we can essentially reduce things to this case: If $f \in \mathbf{M}(I)$ is such that $I-Z(f)$ is uncountable then by Proposition 7.9 there exists a reduction (ψ,g) of f with $\mathrm{supp}(\psi) = \kappa(I-Z(f))$, and $g \in \mathbf{M}_0(I)$. Now most of the important information about the iterates of f is still contained in g , and this information can be "lifted up" to f . If $I-Z(f)$ is countable then we cannot apply Proposition 7.9; but in this case f is very special, and such mappings will be dealt with in Section 9.

We show in Theorem 8.10 that $\mathbf{D}(f)$ is countable for each $f \in \mathbf{M}_0(I)$, and that $\mathbf{D}(f)$ is finite if and only if each f-register-shift is tame. In particular, $\mathbf{D}(f)$ is finite for each $f \in \mathbf{M}_*(I)$, where $\mathbf{M}_*(I)$ denotes the set of mappings $f \in \mathbf{M}_0(I)$ for which there are no f-register-shifts.

Let $f \in \mathbf{M}_*(I)$; it turns out that the maximal elements of the set $\{ D \in \mathbf{D}(f) : D \neq I \}$ are exactly the sets of the form $\kappa(I-A(C,f))$ with C a topologically transitive f-cycle. We call a reduction (ψ,g) of f **primary** if $\mathrm{supp}(\psi) = \kappa(I-A(C,f))$ for some topologically transitive f-cycle C . In Theorem 8.8 we show that if (ψ,g) is a reduction of $f \in \mathbf{M}_*(I)$ with $\mathrm{supp}(\psi) \neq I$ then there exists $n \geq 1$ and $(\psi_1,f_1),\ldots,(\psi_n,f_n)$ such that (ψ_k,f_k) is a primary reduction of f_{k-1} for $k = 1,\ldots,n$ (with $f_0 = f$) and $(\psi,g) = (\psi_n \circ \cdots \circ \psi_1, f_n)$. If $D \in \mathbf{D}(f)$ then by Theorem 7.4 $D = \mathrm{supp}(\psi)$ for some reduction (ψ,g) of f ; thus Theorem 8.8 gives a procedure for constructing the elements of $\mathbf{D}(f)$ for a mapping $f \in \mathbf{M}_*(I)$. It will follow from Theorem 8.8 that if $f \in \mathbf{M}_0(I)$ with $|T(f)| = 1$ then $\mathbf{D}(f)$ is linearly ordered by inclusion.

In Theorem 8.13 we extend Theorem 8.8 to mappings in $\mathbf{M}_0(I)$. For mappings in $\mathbf{M}_0(I)$ it is necessary to use two types of reductions to obtain the analogue of Theorem 8.8: as well as primary reductions (as defined above) we also need what will be called register-shift reductions.

Section 9: Countable closed invariant sets For $f \in \mathbf{M}(I)$ let $\mathbf{I}(f)$ denote the set of closed subsets D of I such that D is f-invariant and $I-D$ is f-almost-invariant (so $D(f) \subset \mathbf{I}(f)$). In this section we study the countable elements in $\mathbf{I}(f)$; the analysis given here is based on ideas to be found in Block (1977), (1979) and Misiurewicz (1980).

Note that if $f \in \mathbf{M}(I)$ and $I-Z(f)$ is countable then $I-Z_*(f)$ is also countable, and it is easy to see that $I-Z_*(I) \in \mathbf{I}(f)$. Theorem 9.5 deals this case, and shows that if $I-Z(f)$ is countable then there exists $p \geq 0$ such that for each $x \in I$ the sequence $\{f^{2^p n}(x)\}_{n\geq 1}$ converges to a periodic point of f ; it follows from this in particular that each periodic point of f has a period which divides 2^p .

We also obtain a similar result for the case when we have f-cycles C' and C with $C' \subset C$ and $C - A(C',f)$ countable. Theorem 9.6 shows that there then exist p , $q \geq 0$ such that $per(C') = 2^p\, per(C)$ and each periodic point in $C - A(C',f)$ has a period which divides $2^q\, per(C)$. Moreover, each point in $C - A(C',f)$ is eventually periodic. ($x \in I$ is eventually periodic if $f^n(x) \in Per(f)$ for some $n \geq 0$.) This result is then used to analyse the structure of tame register-shifts. Let R be an f-register-shift and K be an f-cycle with $R \subset K$ such that $K - A(R,f)$ is countable (i.e. R is tame). Theorem 9.7 then shows that

$$K - A(R,f) = \{ x \in K : x \text{ is eventually periodic } \},$$

and if $x \in K \cap Per(f)$ then x has period $2^j\, per(K)$ for some $j \geq 0$. Moreover, if $\{K_n\}_{n\geq 1}$ is a generator for R with $K_1 \subset K$ then there exists a strictly increasing sequence $\{q_n\}_{n\geq 1}$ of non- negative integers such that $per(K_n) = 2^{q_n}\, per(K)$ for all $n \geq 1$.

Section 10: Extensions Let $f \in \mathbf{M}(I)$ and $\psi \in \mathbf{V}(I)$; we say that $[\psi,f]$ is an **extension** of $g \in \mathbf{M}(I)$ if $\psi \circ f = g \circ \psi$; i.e. $[\psi,f]$ is an extension of g when (ψ,g) is a reduction of f . In this section we study the extensions of a mapping $g \in \mathbf{M}(I)$.

For $g \in \mathbf{M}(I)$ we let $\mathbf{E}(g)$ denote the set of countable subsets E of I such that E is g-invariant and $I-E$ is g-almost-invariant. Let $g \in \mathbf{M}(I)$ and $E \subset I$; in Theorem 10.5 we show that there exists an extension $[\psi,f]$ of g with $\psi(I\text{-supp}(\psi)) = E$ if and only if $E \in \mathbf{E}(g)$. (Note that $\psi(I\text{-supp}(\psi))$ is countable for each $\psi \in V(I)$.)

The main application of Theorem 10.5 is to analyse the mappings in $\mathbf{M}_*(I)$; this is really the converse of the analysis of $\mathbf{M}_*(I)$ given in Section 8. Let $f \in \mathbf{M}(I)$ and C be an f-cycle; we call C **essentially transitive** if there exists a topologically transitive f-cycle $C' \subset C$ such that $C - A(C',f)$ is countable. We call an extension $[\psi,f]$ of $g \in \mathbf{M}(I)$ **primary** if there exists $z \in Per(m,g)$ such that $\psi^{-1}(\{z,g(z),\ldots,g^{m-1}(z)\})$ is an essentially transitive f-cycle and

$$\psi(I - supp(\psi)) \subset \{ y \in I : g^n(y) = z \text{ for some } n \geq 0 \} .$$

Proposition 10.10 shows that if $f \in \mathbf{M}_*(I)$ and (ψ,g) is a primary reduction of f then $[\psi,f]$ is a primary extension of g , and conversely, if $g \in \mathbf{M}_*(I)$ and $[\psi,f]$ is a primary extension of g then $f \in \mathbf{M}_*(I)$ and (ψ,g) is a primary reduction of f . We say that $f \in \mathbf{M}(I)$ is **essentially transitive** if there exists a topologically transitive f-cycle C such that $I - A(C,f)$ is countable. Theorem 10.12 states that if $f \in \mathbf{M}_*(I)$ is not essentially transitive then there exists an essentially transitive mapping $f_0 \in \mathbf{M}_*(I)$, $n \geq 1$ and $[\psi_1,f_1],\ldots, [\psi_n,f_n]$ such that $[\psi_k,f_k]$ is a primary extension of f_{k-1} for $k = 1,\ldots, n$ and $f = f_n$. We can think of Theorem 10.12 as a "construction kit" for building the mappings in $\mathbf{M}_*(I)$. If $f \in \mathbf{M}(I)$ is essentially transitive then it follows from the results in Section 6 and 8 that f is conjugate to a uniformly piecewise linear mapping.

Section 11: Refinements In this section we apply the results of Section 8 to give a fairly complete analysis of the iterates of a mapping $f \in \mathbf{M}_o(I)$. The analysis will be based on the following construct: Let $f \in \mathbf{M}_o(I)$ and $C, C_1,\ldots, C_m$ be f-cycles; we say that $\{C_1,\ldots,C_m\}$ is a **refinement** of C if $int(C_1),\ldots, int(C_m)$ are disjoint, $\bigcup_{j=1}^{m} C_j \subset C$ and $int(C - \bigcup_{j=1}^{m} A(C_j,f)) = \emptyset$. Now given an f-cycle C we will try to find a refinement $\{C_1,\ldots,C_m\}$ of C so that the behaviour of f on the f-invariant closed set $C \quad \bigcup_{j=1}^{m} A(C_j,f)$ is simple enough to be easily described. If this can be done then the problem of studying the behaviour of f on C is really reduced to studying the behaviour of f on the f-cycles $C_1,\ldots, C_m$. We can then repeat this procedure and look for suitable refinements of the f-cycles $C_1,\ldots, C_m$. Our goal is to start this process with the f-cycle I , and after finitely many steps to end up with f-cycles $K_1,\ldots,K_p$ on which the behaviour of f can be directly described (for example, each K_i should be either topologically transitive or contain a single f-register-shift).

This progamme will be realized in Section 11 using just two kinds of refinements. Let $f \in M_0(I)$ and $\{C_1,\ldots,C_m\}$ be a refinement of an f-cycle C ; we say that $\{C_1,\ldots,C_m\}$ is **elementary** (resp. **non-elementary**) if $C - \bigcup_{j=1}^{m} A(C_j,f)$ is countable (resp. uncountable). If $\{C_1,\ldots,C_m\}$ is elementary then Theorem 9.6 describes the behaviour of f on $D = C - \bigcup_{j=1}^{m} A(C_j,f)$: We have that each point in D is eventually periodic, and there exists $p \geq 0$ such that each periodic point of f in D has a period which divides $2^p\,\mathrm{per}(C)$.

Consider next a non-elementary refinement $\{C_1,\ldots,C_m\}$ of an an f-cycle C ; put $D = I - \bigcup_{j=1}^{m} A(C_j,f)$. Then D is closed and f-invariant and $I-D$ is f-almost-invariant; also D is uncountable, and thus by the results of Section 7 there exists a reduction (ψ,g) of f with $\mathrm{supp}(\psi) = \kappa(D)$. Now the restriction of g to $\psi(C)$ essentially describes the restriction of f to $C - \bigcup_{j=1}^{m} A(C_j,f)$, and it follows from a result in Section 7 (Theorem 7.8) that $\psi(C)$ is a g-cycle. Thus the non-elementary refinement $\{C_1,\ldots,C_m\}$ is of the simplest kind when the g-cycle $\psi(C)$ is as simple as possible, and we show that this is the case when $\psi(C)$ is topologically transitive. If $\psi(C)$ is a topologically transitive g-cycle then we call the refinement $\{C_1,\ldots,C_m\}$ **transitive**. The programme outlined above will be carried out using only elementary and transitive refinements.

Section 12: Mappings with one turning point In this section we analyse the structure of a mapping $f \in M_0(I)$ with one turning point. It is easy to see that, without loss of generality, we can restrict ourselves to considering mappings from $\mathbf{S}$, where $\mathbf{S}$ denotes the set of mappings $f \in M_0([0,1])$ having exactly one turning point and for which $f(0) = f(1) = 0$. (Note then that f takes on a maximum at its turning point.) Let $f \in \mathbf{S}$; then by Theorem 2.4 we have that exactly one of the following holds:

(1) There exists a single topologically transitive f-cycle C ; in this case $A(C,f)$ is dense in $[0,1]$.

(2) There exists a single f-register-shift R, and in this case $A(R,f)$ is dense in I.

Let S_t (resp. S_r) denote the set of mappings $f \in S$ for which *(1)* (resp. *(2)*) holds.

We start by considering the mappings in S_t, and introduce a procedure for constructing each element of S_t out of finitely many uniformly piecewise linear mappings. This procedure is based on the following definition: Let $g \in S_t$, and suppose we have an extension $[\psi,f]$ of g with $f \in S$ and such that $\mathrm{supp}(\psi) \neq [0,1]$. Then by a result in Section 8 (Proposition 8.15) the turning point γ of g is periodic, and if m is the period of γ then $B = \psi^{-1}(\{\gamma\})$ is a component of an f-cycle with period m. We thus have $f^m(B) \subset B$, and we show that the restriction of f^m to B is conjugate to an element $q \in S$. In this situation we call $[\psi,f]$ a q **extension** of g. If the turning point of q is periodic then it easily follows that the turning point of f is also periodic; in this case we cannot have an f-register-shift, and so $f \in S_t$.

Let $g \in S_t$ with a periodic turning point, and let $q \in S$; then Proposition 12.1 implies that there exists a q extension of g. Moreover, Proposition 12.2 implies that this extension is uniquely determined (up to conjugacy) by g and q. More precisely, let $g, g_1 \in S_t$, $q, q_1 \in S$ with g and g_1 conjugate and q and q_1 conjugate. Suppose that the turning point of g (and thus also of g_1) is periodic, and let $[\psi,f]$ be a q extension of g and $[\psi_1,f_1]$ be a q_1 extension of g_1. Then f and f_1 are conjugate.

Now let $n \geq 0$ and $h_0, h_1,\ldots, h_n \in S_t$ be essentially transitive, and suppose that for each $k \neq n$ the turning point of h_k is periodic. Then by Proposition 12.1 we can define $f_0,\ldots, f_n \in S_t$ by letting f_0 be conjugate to h_0 and for $k = 1,\ldots, n$ letting f_k be an h_k extension of f_{k-1}. Let $f \in S_t$ be conjugate to f_n; we say then that $h_0,\ldots, h_n$ is an **extension sequence** for f. If $h_0,\ldots, h_n$ (resp. $h'_0,\ldots, h'_n$) is an extension sequence for $f \in S_t$ (resp. $f' \in S_t$) and h_k and h'_k are conjugate for each $k = 0,\ldots, n$ then by Proposition 12.2 f and f' are conjugate. In Theorems 12.3 and 12.4 we show that each element of S_t possesses an (up to conjugacy) unique extension sequence; i.e. if $f \in S_t$ then there exists an extension sequence $h_0,\ldots, h_n$ for f, and if

$h'_0, \ldots, h'_m$ is another extension sequence for f then $m = n$ and h_k and h'_k are conjugate for each $k = 0, \ldots, n$.

For $\beta \in (1,2]$ let u_β be the unique element of **S** which is uniformly piecewise linear with slope β, thus in fact

$$u_\beta(x) = \begin{cases} \beta x & \text{if } 0 \le x \le \frac{1}{2}, \\ \beta - \beta x & \text{if } \frac{1}{2} \le x \le 1. \end{cases}$$

By the result at the end of Section 6 we have $h(u_\beta) = \log \beta$ for each $\beta \in (1,2]$. In particular, if $\alpha, \beta \in (1,2]$ with $\alpha \neq \beta$ then u_α and u_β are not conjugate. (If $f, g \in \mathbf{M}(I)$ are conjugate then $\ell(f^n) = \ell(g^n)$ for all $n \ge 1$, and so $h(f) = h(g)$.)

Let $f \in \mathbf{S}_t$ be essentially transitive; then by the results of Sections 6 and 8 there exists a homeomorphism $\theta \in \mathbf{V}(I)$ and a uniformly piecewise linear mapping $g \in \mathbf{M}(I)$ with slope $\beta = \exp(h(f)) > 1$ such that $\theta \circ f = g \circ \theta$. It follows that $g = u_\beta$, since $g \in \mathbf{S}$; also it is clear that $\beta \le 2$. Thus if $f \in \mathbf{S}_t$ is essentially transitive then there exists a unique $\beta \in (1,2]$ such that f and u_β are conjugate. In Proposition 12.5 we show that, conversely, the mappings u_β, $1 < \beta \le 2$, are all essentially transitive.

From Theorems 12.3 and 12.4 we now have that for each $f \in \mathbf{S}_t$ there exists a unique finite sequence $\beta_0, \ldots, \beta_n$, with $n \ge 0$ and $\beta_k \in (1,2]$ for each $k = 0, \ldots, n$, such that $u_{\beta_0}, \ldots, u_{\beta_n}$ is an extension sequence for f. We call $\beta_0, \ldots, \beta_n$ the **characteristic sequence** of f. It then follows from Proposition 12.2 that two elements of $\mathbf{S}_t$ are conjugate if and only if they have the same characteristic sequence.

Let $\mathbf{P} = \{ \beta \in (1,2] :$ the turning point of u_β is periodic $\}$. If $\beta_0, \ldots, \beta_n$ is the characteristic sequence of some element of $\mathbf{S}_t$ then we must have $\beta_k \in \mathbf{P}$ for each $k \neq n$. On the other hand, if $\beta_0, \ldots, \beta_n$ is any sequence from $(1,2]$ with $\beta_k \in \mathbf{P}$ for each $k \neq n$ then by Propositions 12.1 and 12.5 $\beta_0, \ldots, \beta_n$ is the characteristic sequence of some element of $\mathbf{S}_t$. In Proposition 12.6 we show that $\mathbf{P}$ is countable, and that in fact each $\beta \in \mathbf{P}$ is an algebraic number (i.e. β is a root of some polynomial with integer coefficients).

At the end of Section 12 we extend this analysis to the mappings in $\mathbf{S}_r$. To each mapping in $\mathbf{S}_r$ we associate a characteristic sequence, and show that two mappings are conjugate if and only if they have the same characteristic sequence. If $f \in \mathbf{S}_r$, and the single f-register-shift is tame, then the characteristic sequence of f has the form $\beta_0, \beta_1, \ldots, \beta_p, 1$ with $p \geq 0$ and $\beta_k \in \mathbf{P}$ for each k . If the single f-register-shift is not tame then the characteristic sequence is infinite, and of the form $\beta_0, \beta_1, \beta_2, \ldots$ with $\beta_k \in \mathbf{P}$ for all $k \geq 0$.

Section 13: Some miscellaneous results from real analysis As we have already mentioned, in this section we state and prove some fairly standard results from real analysis which we have used, but which probably do not occur in a typical introductory course. One such result is a simple version of the Baire category theorem, which in particular implies that a residual subset of I is dense.

2. PIECEWISE MONOTONE MAPPINGS

Let a , $b \in \mathbb{R}$ with $a < b$ and put $I = [a,b]$. **C(I)** will denote the set of continuous mappings $f : I \to I$ which map the closed interval I back into itself. If $f \in \mathbf{C}(I)$ and $n \geq 0$ then f^n will denote the **n th. iterate** of f , i.e. $f^n \in \mathbf{C}(I)$ is defined inductively by $f^0(x) = x$, $f^1(x) = f(x)$ and $f^n(x) = f(f^{n-1}(x))$.

We say that $f \in \mathbf{C}(I)$ is **piecewise monotone** if there exists $N \geq 0$ and $a = d_0 < d_1 < \cdots < d_N < d_{N+1} = b$ such that f is strictly monotone on each of the intervals $[d_k, d_{k+1}]$, $k = 0,\ldots, N$. Let f be piecewise monotone and let us make the minimal choice for the d_k's , i.e. so that for $1 \leq k \leq N$ we have f is not monotone in any neighbourhood of d_k ; we then call $d_1,\ldots, d_N$ the **turning points** of f and the intervals $[d_k, d_{k+1}]$, $k = 0,\ldots, N$ the **laps** of f .

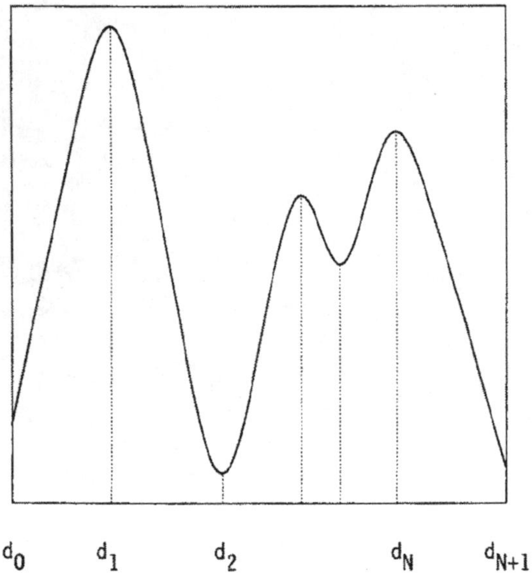

M(I) will denote the set of piecewise monotone mappings in **C(I)** , and for $f \in \mathbf{M}(I)$ let $T(f)$ denote the set of turning points of f . Note that **M(I)** is closed under composition: if f , $g \in \mathbf{M}(I)$ then also $f \circ g \in \mathbf{M}(I)$. This follows because $f \circ g$ is strictly monotone on any interval which does not intersect the finite set $T(g) \cup g^{-1}(T(f))$ (and in fact we have $T(f \circ g) = (T(g) \cup g^{-1}(T(f))) \cap (a,b)$). Thus in particular $f^n \in M(I)$ for any $f \in M(I)$, $n \geq 1$. Moreover, it is easily checked that if $f \in M(I)$ and $n \geq 1$ then

(2.1) $$T(f^n) = \{ x \in (a,b) : f^k(x) \in T(f) \text{ for some } 0 \leq k < n \} .$$

(Let $x \in (a,b)$; then $x \notin T(f^n)$ if and only if f^n is strictly monotone in some neighbourhood of x , and this occurs if and only if for each $k = 0,\ldots, n-1$ we have $f^k(x) \in (a,b)$ and f is strictly monotone in some neighbourhood of $f^k(x)$.)

For $f \in C(I)$ and $n \geq 1$ let $Per(n,f)$ denote the set of **periodic points** of f with **period** n , i.e.

$$Per(n,f) = \{ x \in I : f^n(x) = x , \ f^k(x) \neq x \text{ for } k = 1,\ldots, n-1 \} ;$$

in particular we have $Per(1,f) = Fix(f)$, the set of **fixed points** of f . Note that $Per(n,f) \subset Fix(f^n)$ for each $n \geq 1$. Put $Per(f) = \bigcup_{n\geq 1} Per(n,f)$.

In this section we study the asymptotic behaviour of the iterates of a mapping $f \in M(I)$. The first main result (Theorem 2.4) gives us information about the asymptotic properties of $\{f^n(x)\}_{n\geq 0}$ for a "typical" point $x \in I$. Here "typical" is meant in a topological (rather than in a measure-theoretical) sense: Theorem 2.4 is a statement about the iterates of points from a residual subset of I . ($D \subset I$ is called **residual** if there exists a sequence $\{U_n\}_{n\geq 1}$ of dense open subsets of I with $D \supset \bigcap_{n\geq 1} U_n$. By the Baire category theorem a residual subset of I is dense; see Theorem 13.1 if necessary.)

Theorem 2.4 says that one of three things happens to a "typical" point of I under the iterates of f : Such a point either *(i)* remains in an open set $Z(f)$ on which all the iterates of f are monotone, or *(ii)* ends up in a topologically transitive f-cycle, or *(iii)* gets attracted to an f-register-shift. We must thus now define these objects.

For $f \in M(I)$ let

$$Z(f) = \{ x \in (a,b) : \text{there exists } \varepsilon > 0 \text{ such that } f^n \text{ is monotone on } (x-\varepsilon,x+\varepsilon) \text{ for all } n \geq 0 \} ;$$

clearly $Z(f)$ is open, and in fact by (2.1) $Z(f)$ is the largest open set $G \subset (a,b)$ such that $f^n(G) \cap T(f) = \emptyset$ for all $n \geq 0$.

Lemma 2.1 $f(Z(f)) \subset Z(f)$ for each $f \in \mathbf{M}(I)$.

Proof This is clear, since if f^n is monotone on $(x-\varepsilon, x+\varepsilon)$ for all $n \geq 0$ then $J = f((x-\varepsilon, x+\varepsilon))$ is an open interval with $J \subset (a,b)$ and f^n is monotone on J for all $n \geq 0$. □

We will study the set $Z(f)$ in Section 4.

Let $f \in \mathbf{M}(I)$ and $m \geq 1$; we call a closed set $C \subset I$ an f-**cycle** with **period** m if C is the disjoint union of non-trivial closed intervals $B_0, \ldots, B_{m-1}$ such that $f(B_{k-1}) \subset B_k$ for $k = 1, \ldots, m-1$ and $f(B_{m-1}) \subset B_0$; (in particular we then have $f(C) \subset C$). We call C **proper** if $f(B_{k-1}) = B_k$ for $k = 1, \ldots, m-1$ and $f(B_{m-1}) = B_0$. An f-cycle C is said to be **topologically transitive** if whenever F is a closed subset of C with $f(F) \subset F$ then either $F = C$ or $\text{int}(F) = \emptyset$. Note that a topologically transitive f-cycle is automatically proper.

Remark: There are several other definitions of "topologically transitive" which are equivalent to the one we have used; some of these are listed in Proposition 2.8. (For example, an f-cycle C is topologically transitive if and only if $\overline{\{ f^n(x) : n \geq 0 \}} = C$ for some $x \in C$.)

The simplest example of a mapping having a topologically transitive cycle is probably the "tent map" $h \in \mathbf{M}([0,1])$ (pictured on the next page). Here we have

$$h(x) = \begin{cases} 2x & \text{if } 0 \leq x \leq \frac{1}{2} , \\ 2-2x & \text{if } \frac{1}{2} \leq x \leq 1 , \end{cases}$$

and the whole interval $[0,1]$ is a topologically transitive h-cycle (with period 1). To see this consider a non-trivial interval $J \subset [0,1]$. If $\frac{1}{2} \notin h^n(J)$ then the interval $h^{n+1}(J)$ is twice as long as $h^n(J)$, and hence we must have $\frac{1}{2} \in h^m(J)$ for some $m \geq 0$. But then $0 \in h^{m+2}(J)$ and so $0 \in h^n(J)$ for all $n \geq m+2$. The same argument now shows that $\frac{1}{2} \in h^p(J)$ for some $p \geq m+2$ and thus

$h^{p+1}(J) = I$. This shows that if F is a closed subset of $[0,1]$ with $f(F) \subset F$ and $\mathrm{int}(F) \neq \emptyset$ then $F = [0,1]$.

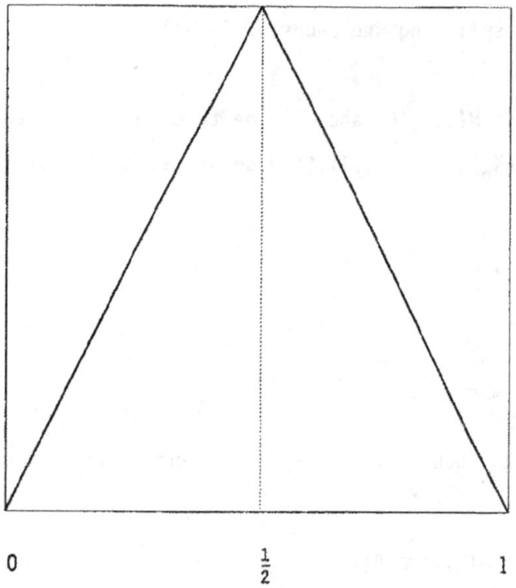

Let $f \in M(I)$ and C be an f-cycle; we let $\mathrm{per}(C)$ denote the period of the f-cycle C . If C_1 and C_2 are f-cycles with $C_2 \subset C_1$ then it is easy to see that $\mathrm{per}(C_1)|\mathrm{per}(C_2)$ and that each component of C_1 contains $\mathrm{per}(C_2)/\mathrm{per}(C_1)$ components of C_2 . Let $\{C_n\}_{n\geq 1}$ be a decreasing sequence of f-cycles, C_n having period m_n ; we call the sequence $\{C_n\}_{n\geq 1}$ **splitting** if $m_{n+1} > m_n$ for each $n \geq 1$.

We say that $R \subset I$ is an f-**register-shift** if there exists a splitting sequence of proper f-cycles $\{K_n\}_{n\geq 1}$ with $K_1 \cap Z(f) = \emptyset$ such that $R = \bigcap_{n\geq 1} K_n$, and we then call $\{K_n\}_{n\geq 1}$ a **generator** for R . If R is an f-register-shift then Propositions 2.2 and 2.10 will show that $\mathrm{int}(R) = \emptyset$ and that f maps R homeomorphically back onto itself; moreover, the sequence $\{f^n(x)\}_{n\geq 0}$ is dense in R for each $x \in R$. In Section 5 we give some simple examples to show how register-shifts can occur.

If C is an f-cycle then let

$$A(C,f) = \{ x \in I : f^n(x) \in \mathrm{int}(C) \ \text{ for some } \ n \geq 0 \} ;$$

and put $A(\{C_n\}_{n\geq 1},f) = \bigcap_{n\geq 1} A(C_n,f)$ when $\{C_n\}_{n\geq 1}$ is a splitting sequence of f-cycles. The following result lists some elementary properties of topologically transitive f-cycles and splitting sequences of f-cycles.

Proposition 2.2 Let $f \in M(I)$, C and C' be topologically transitive f-cycles, and let $\{K_n\}_{n\geq 1}$ and $\{K'_n\}_{n\geq 1}$ be splitting sequences of proper f-cycles. Then:

(1) $T(f) \cap \mathrm{int}(C) \neq \emptyset$.

(2) $T(f) \cap \bigcap_{n\geq 1} \mathrm{int}(K_n) \neq \emptyset$.

(3) $A(C,f) \cap Z(f) = \emptyset$.

(4) $\mathrm{int}(\bigcap_{n\geq 1} K_n - Z(f)) = \emptyset$.

(5) There exists $m \geq 1$ such that $C \cap K_m = \emptyset$, and so in particular $A(C,f) \cap A(\{K_n\}_{n\geq 1},f) = \emptyset$.

(6) Either $C = C'$ or $A(C,f) \cap A(C',f) = \emptyset$.

(7) Either $\bigcap_{n\geq 1} K_n = \bigcap_{n\geq 1} K'_n$ and $A(\{K_n\}_{n\geq 1},f) = A(\{K'_n\}_{n\geq 1},f)$ or $(\bigcap_{n\geq 1} K_n) \cap (\bigcap_{n\geq 1} K'_n) = \emptyset$ and $A(\{K_n\}_{n\geq 1},f) \cap A(\{K'_n\}_{n\geq 1},f) = \emptyset$.

Proof This is very straightforward, and is given at the end of the section. □

If R is an f-register-shift then let $A(R,f) = A(\{K_n\}_{n\geq 1},f)$, where $\{K_n\}_{n\geq 1}$ is any generator for R ; by Proposition 2.2 (7) this is well-defined. We will later see (in Proposition 2.9) that in fact

$$A(R,f) = \{ x \in I : \lim_{n\to\infty} \min_{z\in R} |f^n(x)-z| = 0 \} .$$

If R is an f-register-shift then Proposition 2.2 (4) implies that $\mathrm{int}(R) = \emptyset$.

Proposition 2.3 Let $f \in M(I)$, let $C_1,\ldots, C_r$ be the topologically transitive f-cycles and let $R_1,\ldots, R_\ell$ be the f-register-shifts. Then $\ell+r \leq |T(f)|$ and the sets $A(C_1,f),\ldots, A(C_r,f)$, $A(R_1,f),\ldots, A(R_\ell,f)$ and $Z(f)$ are disjoint.

Proof This follows immediately from Proposition 2.2. □

We now come to the first main result of this section.

Theorem 2.4 Let $f \in M(I)$, let $C_1, \dots, C_r$ be the topologically transitive f-cycles and let $R_1, \dots, R_\ell$ be the f-register-shifts. Then

$$\Lambda(f) = A(C_1,f) \cup \cdots \cup A(C_r,f) \cup A(R_1,f) \cup \cdots \cup A(R_\ell,f) \cup Z(f)$$

is a dense subset of I.

Proof In Section 3. □

Remark: $\Lambda(f)$ can clearly be written as a countable intersection of open subsets of I, and thus Theorem 2.4 implies that $\Lambda(f)$ is residual.

Consider the special situation where we have $f \in M(I)$ with $|T(f)| = 1$; for such a mapping Theorem 2.4 tells us that exactly one of the following holds:

(2.2) There is a topologically transitive f-cycle C; in this case $A(C,f) \cup Z(f)$ is dense in I.

(2.3) There is an f-register-shift R; in this case $A(R,f) \cup Z(f)$ is dense in I.

(2.4) $Z(f)$ is dense in I.

The second main result of this section (Theorem 2.5) gives us more information about the topologically transitive f-cycles. We say that $f \in M(I)$ is **(topologically) exact** if for each non-trivial interval $J \subset I$ there exists $n \geq 0$ such that $f^n(J) = I$. We call $f \in M(I)$ **semi-exact** if there exists $c \in (a,b)$ such that $f([a,c]) = [c,b]$, $f([c,b]) = [a,c]$, and the restriction of f^2 to $[a,c]$ is exact. We have in fact already shown that the "tent map" is exact. A simple example of a semi-exact mapping is pictured on the following page.

Now let $f \in M(I)$ and C be a proper f-cycle with period m; let B be one of the m components of C. We say that C is **exact** (resp. **semi-exact**) if the restriction of f^m to B is exact (resp. semi-exact). (It is easy to see that these definitions do not depend on which of the components of C is used.) If C

is either exact or semi-exact then C is clearly topologically transitive, and the next result shows that the converse of this is true.

Theorem 2.5 Let $f \in \mathbf{M}(I)$ and C be a topologically transitive f-cycle. Then C is either exact or semi-exact.

Proof Also in Section 3. □

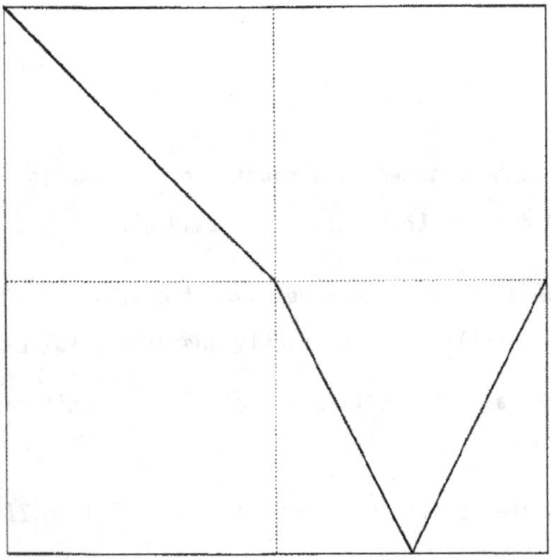

Let $f \in \mathbf{M}(I)$ and C be a topologically transitive f-cycle with period m ; let B be one of the m components of C and g be the restriction of f^m to B . Then the asymptotic behaviour of the iterates of f on the set $A(C,f)$ is determined by the properties of the mapping $g \in \mathbf{M}(B)$. But by Theorem 2.5 C is either exact or semi-exact, and so g is also either exact or semi-exact. We now give a couple of important facts about such mappings. Let f , $g \in \mathbf{C}(I)$; f and g are said to be **conjugate** if there exists a homeomorphism $\psi : I \to I$ (i.e. ψ is a continuous and strictly monotone mapping of I onto itself) such that $\psi \circ f = g \circ \psi$. Let $f \in \mathbf{M}(I)$ and $\beta > 0$; f is said to be **uniformly piecewise linear** with **slope** β if on each of its laps f is linear with slope β or $-\beta$.

Theorem 2.6 Let $f \in \mathbf{M}(I)$ be exact or semi-exact. Then f is conjugate to a uniformly piecewise linear mapping.

Proof This is a result in Parry (1966): $g \in \mathbf{M}(I)$ is called **strongly transitive** if for each non-trivial interval $J \subset I$ there exists $n \geq 0$ such that $I = \bigcup_{k=0}^{n} g^k(J)$. If $f \in \mathbf{M}(I)$ is either exact or semi-exact then clearly it is strongly transitive. Corollary 3 in Parry (1966) states that a strongly transitive mapping $f \in \mathbf{M}(I)$ is conjugate to a uniformly piecewise linear mapping. □

In Section 6 we give another proof of Theorem 2.6, using a method developed in Milnor and Thurston (1977).

For $f \in \mathbf{M}(I)$ and $\varepsilon > 0$ let

$$\Sigma_\varepsilon(f) = \{\, x \in I : \text{for each } \delta > 0 \text{ there exists } n \geq 0 \text{ such that } |f^k((x-\delta,x+\delta))| \geq \varepsilon \text{ for all } k \geq n \,\},$$

(where if J is an interval then $|J|$ denotes the length of J). It is easy to see that $\Sigma_\varepsilon(f)$ is a closed subset of I . We say that f has **sensitive dependence to initial conditions** if $\Sigma_\varepsilon(f) = I$ for some $\varepsilon > 0$.

Proposition 2.7 Let $f \in \mathbf{M}(I)$ be either exact or semi-exact. Then:

(1) f has sensitive dependence to initial conditions.

(2) $\mathrm{Per}(f)$ is dense in I .

Proof (1): If f is exact then clearly we have $\Sigma_\varepsilon(f) = I$ with $\varepsilon = |I|$. Similarly, if f is semi-exact then $\Sigma_\varepsilon(f) = I$ holds with $\varepsilon = \min\{c-a, b-c\}$, where $c \in (a,b)$ is as in the definition of f being semi-exact.

(2): Suppose f is exact and let $J \subset I$ be a non-trivial closed interval. Then there exists $n \geq 0$ such that $f^n(J) = I$ and hence $f^n(x) = x$ for some $x \in J$. Thus $J \cap \mathrm{Per}(f) \neq \emptyset$ and so $\mathrm{Per}(f)$ is dense in I . If f is semi-exact then the same proof shows that $\mathrm{Per}(f^2)$ is dense in I (since $f(\mathrm{Per}(f^2)) \subset \mathrm{Per}(f^2)$), and therefore $\mathrm{Per}(f)$ is also dense in I . □

Let $f \in M(I)$ and C be a topologically transitive f-cycle. It follows immediately from Proposition 2.7 and Theorem 2.5 that $Per(f) \cap C$ is dense in C and $A(C,f) \subset \sum_{\varepsilon}(f)$ for some $\varepsilon > 0$.

Let $f \in M(I)$ and C be an f-cycle; the next result lists some properties which are equivalent to C being topologically transitive. This result is quite standard, and can be found, for example, in Walters (1982). Moreover, it holds quite generally, and does not depend on the particular one-dimensional situation we are considering.

Proposition 2.8 Let $f \in M(I)$ and C be an f-cycle. Then the following are equivalent:

(1) C is topologically transitive.

(2) $\{ x \in C : f^n(x) \in U$ for some $n \geq 0 \}$ is a dense open subset of C for each non-empty open subset U of C .

(3) $\{ x \in C : \overline{\{ f^n(x) : n \geq 0 \}} = C \}$ is a residual subset of C .

(4) $\overline{\{ f^n(x) : n \geq 0 \}} = C$ for some $x \in C$.

Proof This is given at the end of the section. □

We next consider some of the properties of f-register-shifts.

Proposition 2.9 Let $f \in M(I)$ and R be an f-register-shift. Then

$$A(R,f) = \{ x \in I : \lim_{n\to\infty} \min_{z\in R} |f^n(x)-z| = 0 \}$$
$$= \{ x \in I : \liminf_{n\to\infty} \min_{z\in R} |f^n(x)-z| = 0 \} .$$

Thus $x \notin A(R,f)$ if and only if there exists $\varepsilon > 0$ such that $|f^n(x)-z| \geq \varepsilon$ for all $n \geq 0$ and $z \in R$.

Proof Put $D = \{ x \in I : \lim_{n\to\infty} \min_{z\in R} |f^n(x)-z| = 0 \}$ and $D' = \{ x \in I : \liminf_{n\to\infty} \min_{z\in R} |f^n(x)-z| = 0 \}$; clearly $D \subset D'$. Let $\{K_n\}_{n\geq 1}$ be a generator for R , and suppose K_n has period q_n . Now for each $\delta > 0$ there exists $n \geq 1$ such that $\min_{z \in R} |x-z| < \delta$ for all $x \in K_n$, and hence $A(R,f) \subset D$. Conversely, we show that $D' \subset A(K_n,f)$ for each $n \geq 1$. Fix $n \geq 1$; since $q_{n+2} \geq 4q_n$ and each component of K_{n+2} contains q_{n+2}/q_n components of K_{n+2}, there exists a component B of K_{n+2} and $\delta > 0$ such that if $z \in B$ and $|y-z| < \delta$ then $y \in \mathrm{int}(K_n)$. Put $q = q_{n+2}$ and let $\varepsilon > 0$ be such that $\max\{|J|,|f(J)|,\ldots,|f^{q-1}(J)|\} < \delta$ whenever J is an interval with $|J| < \varepsilon$. (ε exists because the mappings $f,\ldots, f^{q-1}$ are uniformly continuous.) Now take $x \in D'$; there thus exists $m \geq 0$ and $z \in R$ such that $|f^m(x)-z| < \varepsilon$, and so $|f^{m+k}(x)-f^k(z)| < \delta$ for $k = 0,\ldots, q-1$. But $f^k(z) \in B$ for some $0 \leq k < q-1$, and therefore $f^{m+k}(x) \in \mathrm{int}(K_n)$ for some $k \geq 0$; hence $x \in A(K_n,f)$ and so $D' \subset A(K_n,f)$. □

Let $f \in \mathbf{M}(I)$ and R be an f-register-shift. If $\{K_n\}_{n\geq 1}$ is a generator for R then $f(K_n) = K_n$ for each $n \geq 1$, and it thus follows that $f(R) = R$ (since $\{K_n\}_{n\geq 1}$ is a decreasing sequence of compact sets).

Proposition 2.10 Let $f \in \mathbf{M}(I)$ and R be an f-register-shift. Then f maps R homeomorphically back onto itself, and $\{f^n(x)\}_{n\geq 0}$ is dense in R for each $x \in R$.

Proof Let $\{K_n\}_{n\geq 1}$ be a generator for R. Suppose there exist x, $y \in R$ with $x < y$ and $f(x)= f(y)$. Then for each $n \geq 1$ we have that x and y are in the same component of K_n (namely in the component which "precedes" the component containing $f(x)$), and thus $[x,y] \subset K_n$. This implies that $[x,y] \subset R$, which contradicts the fact that $\mathrm{int}(R) = \emptyset$. Thus f is injective on R and, since we have already seen that $f(R) = R$, the mapping $f : R \to R$ is therefore a bijection. Hence f maps R homeomorphically back onto itself, i.e. the inverse of f is also continuous; (if this is not clear then see Lemma 13.3). Now let x, $y \in R$ and $\delta > 0$. Since $\mathrm{int}(R) = \emptyset$ there exists $m \geq 1$ such that if B is the component of K_m containing y then $|B| < \delta$. But for some $n \geq 0$ we have $f^n(x) \in B$ and hence $|f^n(x)-y| < \delta$. This shows that $\{f^n(x)\}_{n\geq 0}$ is dense in R. □

The next result shows it was not really necessary to assume that the f-cycles in the definition of an f-register-shift be proper.

Proposition 2.11 Let $f \in M(I)$ and $\{K_n\}_{n\geq 1}$ be a splitting sequence of f-cycles with $K_1 \cap Z(f) = \emptyset$. Then $\bigcap_{n\geq 1} K_n$ is an f-register-shift.

Proof If C is an f-cycle then we put $\hat{C} = \bigcap_{n\geq 0} f^n(C)$.

Lemma 2.12 Let C be an f-cycle with period m. Then either $\hat{C}$ is a proper f-cycle with period m, or $\hat{C} = \{x, f(x), \ldots, f^{m-1}(x)\}$ for some $x \in \mathrm{Per}(m,f)$.

Proof Let $B_0, \ldots, B_{m-1}$ be the m components of C, labelled so that $f(B_{k-1}) \subset B_k$ for $k = 1, \ldots, m-1$ and $f(B_{m-1}) \subset B_0$. It is easy to see that $\hat{C} = \bigcup_{k=0}^{m-1} \hat{B}_k$, where $\hat{B}_k = \bigcap_{n\geq 0} f^{nm}(B_k)$, and that $f(\hat{B}_{k-1}) = \hat{B}_k$ for $k = 1, \ldots, m-1$ and $f(\hat{B}_{m-1}) = \hat{B}_0$. But each $\hat{B}_k$ is a closed interval, thus either $\hat{B}_0 = \{x\}$ for some $x \in I$, in which case $x \in \mathrm{Per}(m,f)$ and $\hat{C} = \{x, f(x), \ldots, f^{m-1}(x)\}$, or all the intervals $\hat{B}_0, \ldots, \hat{B}_{m-1}$ are non-trivial and then $\hat{C}$ is a proper f-cycle with period m. □

Lemma 2.13 Let C be an f-cycle and suppose there exists an f-cycle $D \subset C$ with $\mathrm{per}(D) > \mathrm{per}(C)$. Then $\hat{C}$ is a proper f-cycle.

Proof Clearly $\hat{D} \subset \hat{C}$ and so by Lemma 2.12 we have $|\hat{C}| \geq |\hat{D}| \geq \mathrm{per}(D) > \mathrm{per}(C)$. Hence, again using Lemma 2.12, we have that $\hat{C}$ is a proper f-cycle. □

By Lemmas 2.12 and 2.13 $\{\hat{K}_n\}_{n\geq 1}$ is a splitting sequence of proper f-cycles, and since $\hat{K}_1 \subset K_1$ it follows that $R = \bigcap_{n\geq 1} \hat{K}_n$ is an f-register-shift. We will show that $R = \bigcap_{n\geq 1} K_n$, and for this it is enough to show that $\bigcap_{n\geq 1} K_n \subset \hat{K}_m$ for each $m \geq 1$. Now by Proposition 2.2 (4) we have $\mathrm{int}(\bigcap_{n\geq 1} K_n - Z(f)) = \emptyset$ (since the proof of this part of Proposition 2.2 does not require that the f-cycles $\{K_n\}_{n\geq 1}$ be proper), and hence $\mathrm{int}(\bigcap_{n\geq 1} K_n) = \emptyset$. Therefore if $\bigcap_{n\geq 1} K_n \not\subset \hat{K}_m$ for some $m \geq 1$ then there exists $p > m$ and a component B of K_p such that $B \cap \hat{K}_m = \emptyset$. But this is clearly not possible, and so $\bigcap_{n\geq 1} K_n \subset \hat{K}_m$ for each $m \geq 1$. □

We now come to the proof of Proposition 2.2, and for this we need two general facts which will also be used in later sections.

Lemma 2.14 Let $f \in \mathbf{M}(I)$ and C and C' be f-cycles. Then $A(C,f) \cap A(C',f) \neq \emptyset$ if and only if $\mathrm{int}(C \cap C') \neq \emptyset$.

Proof If $\mathrm{int}(C \cap C') \neq \emptyset$ then clearly $A(C,f) \cap A(C',f) \neq \emptyset$. Conversely, if $A(C,f) \cap A(C',f) \neq \emptyset$ then we can find a non-trivial interval $J \subset A(C,f) \cap A(C',f)$. Now for each $x \in J$ there exists $m \geq 0$ such that $f^n(x) \in C \cap C'$ for all $n \geq m$. But if $\mathrm{int}(C \cap C') = \emptyset$ then $C \cap C'$ is finite and then we would have that $\{ x \in J : f^n(x) \in C \cap C' \text{ for some } n \geq 0 \}$ is countable; thus, since J is uncountable, we must have $\mathrm{int}(C \cap C') \neq \emptyset$. □

Lemma 2.15 Let $f \in \mathbf{M}(I)$ and C and C' be f-cycles with C proper and $C \cap C' \neq \emptyset$. If $\mathrm{per}(C) > 2\,\mathrm{per}(C')$ then $C \subset C'$.

Proof Suppose that C is not a subset of C' ; let $m = \mathrm{per}(C)$ and $B_0,\dots, B_{m-1}$ be the components of C . If $B_j \subset C'$ for some j then $C \subset C'$ (since C is proper); hence for each k we have that B_k is not a subset of C' . On the other hand we clearly have $B_k \cap C' \neq \emptyset$ for each k , and so $B_k \cap \partial C' \neq \emptyset$ for each k . But $|\partial C'| \leq 2\,\mathrm{per}(C')$, and therefore $m \leq 2\,\mathrm{per}(C')$. □

Proof of Proposition 2.2 (1): Let $m = \mathrm{per}(C)$ and B be one of the m components of C . If $T(f) \cap \mathrm{int}(C) = \emptyset$ then $f^{2m}(B) = B$ and f^{2m} is increasing on B . We can thus find a non-trivial closed interval $J \subset B$ such that $f^{2m}(J) \subset J$ and $J \cup f^m(J) \neq B$. But this would imply that C is not topologically transitive; hence $T(f) \cap \mathrm{int}(C) \neq \emptyset$.

(2): Let $q_n = \mathrm{per}(K_n)$; since $q_{n+1} \geq 2q_n$ we have $\lim_{n\to\infty} q_n = \infty$. Suppose $T(f) \cap \mathrm{int}(K_m) = \emptyset$ for some $m \geq 1$ and let B be one of the components of K_m . Then f^{2q_m} is increasing on B , and thus for each $x \in B$ the sequence $(f^{2nq_m}(x))_{n\geq 1}$ converges to an element of $\mathrm{Fix}(f^{2q_m})$. But for $j > m$ we have $K_j \cap B \neq \emptyset$ and $f(K_j) \subset K_j$; hence $K_j \cap \mathrm{Fix}(f^{2q_m}) \neq \emptyset$. It follows that $q_j | 2q_m$ for all $j > m$, which is of course not possible.

(3): Suppose that $A(C,f) \cap Z(f) \neq \emptyset$; then by Lemma 2.1 we have $\overline{C \cap Z(f)} \neq \emptyset$ and so there exists a non-trivial interval $J \subset C \cap Z(f)$. Let $F = \overline{\bigcup_{n\geq 1} f^n(J)}$; then $f(F) \subset F$ and $\mathrm{int}(F) \neq \emptyset$; thus $F = C$ and in particular $J \cap f^m(J) \neq \emptyset$ for some $m \geq 1$. Hence $L = \bigcup_{n\geq 0} f^{nm}(J)$ is a non-trivial interval with $f^m(L) \subset L$. By Lemma 2.1 $L \subset Z(f)$ and therefore f^m is monotone on L . But we also have $L \subset C$, and as in (1) this contradicts the fact that C is topologically transitive.

(4): Let $J \subset \bigcap_{n\geq 1} K_n$ be a non-trivial open interval. If K_n has period q_n then J must be contained in one of the q_n components of K_n , and hence the intervals $J, f(J),\ldots, f^{q_n-1}(J)$ are disjoint. Since $\lim_{n\to\infty} q_n = \infty$ all the intervals $\{f^n(J)\}_{n\geq 0}$ are disjoint, and so there exists $k \geq 0$ such that $f^k(J) \subset (a,b)$ and $f^n(J) \cap T(f) = \emptyset$ for all $n \geq k$. Then f^n is monotone on $f^k(J)$ for all $n \geq 0$, and therefore $\mathrm{int}(f^k(J)) \subset Z(f)$. It follows that $J - Z(f) \subset \{a,b\} \cup T(f^k)$, which is finite, and so $\mathrm{int}(\bigcap_{n\geq 1} K_n - Z(f)) = \emptyset$.

Note: We did not use the fact that the f-cycles $\{K_n\}_{n\geq 1}$ are proper in this last part of the proof.

(5): Suppose $C \cap K_n \neq \emptyset$ for all $n \geq 1$; then by Lemma 2.15 there exists $m \geq 1$ such that $K_n \subset C$ for all $n \geq m$. This cannot happen because C is topologically transitive.

(6): Suppose $A(C,f) \cap A(C',f) \neq \emptyset$; then by Lemma 2.14 $\mathrm{int}(C \cap C') \neq \emptyset$ and so, putting $L = C \cap C'$, we have that L is closed, $f(L) \subset L$ and $\mathrm{int}(L) \neq \emptyset$. Thus $C = L = C'$.

(7): If $A(\{K_n\}_{n\geq 1},f) \cap A(\{K'_n\}_{n\geq 1},f) \neq \emptyset$ then $K_n \cap K'_m \neq \emptyset$ for all n , $m \geq 1$ and hence $(\bigcap_{n\geq 1} K_n) \cap (\bigcap_{n\geq 1} K'_n) \neq \emptyset$. Suppose that $(\bigcap_{n\geq 1} K_n) \cap (\bigcap_{n\geq 1} K'_n) \neq \emptyset$; then $K_n \cap K'_m \neq \emptyset$ for all n , $m \geq 1$ and so Lemma 2.15 implies that for each $n \geq 1$ there exist p , $q \geq 1$ such that $K'_p \subset K_n$ and $K_q \subset K'_n$. Therefore $\bigcap_{n\geq 1} K_n = \bigcap_{n\geq 1} K'_n$ and also $A(\{K_n\}_{n\geq 1},f) = A(\{K'_n\}_{n\geq 1},f)$. □

Proof of Proposition 2.8 (1) implies (2) : Let U be a non-empty open subset of C , and put $F = C - V$, where $V = \{ x \in C : f^n(x) \in U$ for some $n \geq 0 \}$. Then F is a closed subset of C , and clearly $f(F) \subset F$. Also $F \neq C$, and hence we must have $\text{int}(F) = \emptyset$; therefore V is a dense (open) subset of C .

(2) implies (3) : Let $\{U_m\}_{m\geq 1}$ be a countable basis for the topology on C . (Thus each U_m is a non-empty open subset of C , and for each $x \in C$ and each neighbourhood N of x in C there exists $m \geq 1$ with $x \in U_m \subset N$. It is clear that a sequence $\{U_m\}_{m\geq 1}$ with these properties exists.) Let $x \in C$; then $\overline{\{ f^n(x) : n \geq 0 \}} = C$ if and only if $\{ f^n(x) : n \geq 0 \} \cap U_m \neq \emptyset$ for each $m \geq 1$, and this holds if and only if $x \in V_m$ for each $m \geq 1$, where $V_m = \{ y \in C : f^n(y) \in U_m$ for some $n \geq 0 \}$. This shows that $\{ x \in C : \overline{\{ f^n(x) : n \geq 0 \}} = C \} = \bigcap_{m\geq 1} V_m$. But by assumption each V_m is a dense open subset of C , and hence $\{ x \in C : \overline{\{ f^n(x) : n \geq 0 \}} = C \}$ is a residual subset of C .

(3) implies (4) : This is clear, since by the Baire category theorem (see Theorem 13.1 if necessary) a residual subset of C is dense, and so in particular non-empty.

(4) implies (1) : Let F be a closed subset of C with $f(F) \subset F$ and $\text{int}(F) \neq \emptyset$; let $x \in C$ be such that $\overline{\{ f^n(x) : n \geq 0 \}} = C$. There thus exists $m \geq 0$ with $f^m(x) \in F$, and then $f^n(x) \in F$ for all $n \geq m$, since $f(F) \subset F$. Therefore $\overline{\{ f^n(x) : n \geq m \}} \subset \overline{F} = F$, and so $F = C$, because $\overline{\{ f^n(x) : n \geq m \}} = \overline{\{ f^n(x) : n \geq 0 \}}$. □

Remark: The proof of Proposition 2.8 almost remains valid if C is replaced by an arbitrary compact metric space X , and f by an arbitrary continuous mapping of X into itself. The one problem is in the proof that (4) implies (1), where we implicitly used the fact that C contains no isolated points. In general, (1), (2), (3) and (4)′ are equivalent, where (4)′ is (4) together with the condition that f be onto.

3. PROOF OF THEOREMS 2.4 AND 2.5

Let $f \in M(I)$ with topologically transitive f-cycles $C_1, \ldots, C_r$ and f-register-shifts $R_1, \ldots, R_\ell$; put

$$\Lambda(f) = A(C_1,f) \cup \cdots \cup A(C_r,f) \cup A(R_1,f) \cup \cdots \cup A(R_\ell,f) \cup Z(f) .$$

We show first that $\Lambda(f)$ is a dense subset of I ; i.e. we will show that $\Lambda(f) \cap J \neq \emptyset$ for each non-empty open interval $J \subset I$.

Let $J \subset I$ be a non-empty open interval, and put $U = \mathrm{int}(\bigcup_{n\geq 0} f^n(J))$; thus $J \subset U$. Let U_0 be the maximal connected component of U containing J . Now for each $m \geq 1$ we have $f^m(U_0) \subset \bigcup_{n\geq 0} f^n(J)$ (since $U_0 \subset \bigcup_{n\geq 0} f^n(J)$) and $f^m(U_0)$ is an interval; there thus exists a unique component U_m of U such that $f^m(U_0) \subset \overline{U_m}$. It is easy to see that we then have $f^p(U_q) \subset \overline{U_{p+q}}$ for all $p , q \geq 0$. Note that either $U_n = U_m$ or $U_n \cap U_m = \emptyset$ for each $n , m \geq 0$.

Lemma 3.1 If $U_n \cap U_m = \emptyset$ for all $n > m \geq 0$ then $J \cap Z(f) \neq \emptyset$.

Proof Since $T(f) \cup \{a,b\}$ is finite there exists $m > 0$ such that $U_n \cap (T(f) \cup \{a,b\}) = \emptyset$ for all $n \geq m$, and then we clearly have $U_m \subset Z(f)$. It follows that $U_0 - Z(f) \subset \{a,b\} \cup T(f^m)$, which is finite, and so in particular $J \cap Z(f) \neq \emptyset$. □

We now assume that $U_m = U_n$ for some $n > m \geq 0$, and put $C = \bigcup_{j=m}^{n-1} \overline{U_j}$. Then, since $f(\overline{U_{j-1}}) \subset \overline{U_j}$ for $j = m+1, \ldots, n-1$ and $f(\overline{U_{n-1}}) \subset \overline{U_n} = \overline{U_m}$, the next lemma shows that C is an f-cycle.

Lemma 3.2 Let $f \in M(I)$ and $B_0, \ldots, B_{p-1}$ be non-trivial closed intervals with $f(B_{k-1}) \subset B_k$ for $k = 1, \ldots, p-1$ and $f(B_{p-1}) \subset B_0$. Then $Q = \bigcup_{k=0}^{p-1} B_k$ is an f-cycle.

Proof For $k = 0, \ldots, p-1$ let Q_k be the largest closed interval with $B_k \subset Q_k \subset Q$. Since $f(Q_{k-1})$ is a closed interval with $f(Q_{k-1}) \cap Q_k \neq \emptyset$ we must have $f(Q_{k-1}) \subset Q_k$ for $k = 1,\ldots, p-1$ and $f(Q_{p-1}) \subset Q_0$. Also, if $0 \le j < k < p$ then either $Q_j = Q_k$ or $Q_j \cap Q_k = \emptyset$. Let $q = \min\{ k > 0 : Q_k = Q_0 \}$; it is then easily checked that $Q_0,\ldots, Q_{q-1}$ are disjoint and $Q = \bigcup_{k=0}^{q-1} Q_k$. Hence Q is an f-cycle. □

We continue to let $C = \bigcup_{j=m}^{n-1} \overline{U_j}$.

Lemma 3.3 (1) If C' is an f-cycle with $C' \subset C$ then $J \cap A(C',f) \neq \emptyset$.

(2) If R is an f-register-shift with $R \subset C$ then $J \cap A(R,f) \neq \emptyset$.

(3) If $C \cap Z(f) \neq \emptyset$ then $J \cap Z(f) \neq \emptyset$.

Proof (1): $C - \bigcup_{j=m}^{n-1} U_j$ is finite and $\mathrm{int}(C')$ is infinite, and so $\mathrm{int}(C') \cap \bigcup_{j=m}^{n-1} U_j \neq \emptyset$. There thus exists $x \in J$ and $k \ge 0$ such that $f^k(x) \in \mathrm{int}(C')$; i.e. $x \in J \cap A(C',f)$.

(2): This is the same as (1), because R is infinite and if $f^k(x) \in R$ for some $k \ge 0$ then $x \in A(R,f)$.

(3): Since $C \cap Z(f)$ is infinite there exists $x \in J$ and $k \ge 0$ such that $f^k(x) \in Z(f)$. Thus for some $\varepsilon > 0$ we have either $(x-\varepsilon,x) \subset Z(f)$ or $(x,x+\varepsilon) \subset Z(f)$, and hence $J \cap Z(f) \neq \emptyset$. □

Lemmas 3.1, 3.2 and 3.3 reduce the problem of showing that $\Lambda(f)$ is dense in I to proving the following result:

Theorem 3.4 Let $f \in M(I)$ and C be an f-cycle with $C \cap Z(f) = \emptyset$. Then C contains either an f-register-shift or a topologically transitive f-cycle.

Proof Let $N = \sup\{ \mathrm{per}(C') : C'$ is an f-cycle with $C' \subset C \}$.

Lemma 3.5 If $N = \infty$ then C contains an f-register-shift.

Proof Since $C \cap Z(f) = \emptyset$ we have $\mathrm{int}(C') \cap T(f) \neq \emptyset$ for each f-cycle $C' \subset C$.

There thus exists $w \in \mathrm{int}(C') \cap T(f)$ such that if $\mathbf{B}_w$ is the set of f-cycles $K \subset C$ with $w \in \mathrm{int}(K)$ then $\sup\{ \mathrm{per}(K) : K \in \mathbf{B}_w \} = \infty$. Now if K, $K' \in \mathbf{B}_w$ then there is a $K'' \in \mathbf{B}_w$ with $K'' \subset K \cap K'$. (Let B (resp. B') be the component of K (resp. K') containing w, and let $B'' = B \cap B'$. We have $w \in \mathrm{int}(B'')$ and $f^p(B'') \subset B''$, where p is the least common multiple of $\mathrm{per}(K)$ and $\mathrm{per}(K')$. By Lemma 3.2 $K'' = \bigcup_{j=0}^{p-1} f^j(B'')$ is an f-cycle, and clearly $K'' \in \mathbf{B}_w$ and $K'' \subset K \cap K'$.) Note that we have $\mathrm{per}(K'') \geq \max\{\mathrm{per}(K),\mathrm{per}(K')\}$. Therefore there exists a splitting sequence of f-cycles $(K_n)_{n\geq 1}$ with $K_1 \subset C$, and it follows from Proposition 2.11 that $\bigcap_{n\geq 1} K_n$ is an f-register-shift. □

Let us call an f-cycle K **minimal** if whenever K' is an f-cycle with $K' \subset K$ then $K' = K$.

Lemma 3.6 If $N < \infty$ then C contains a minimal f-cycle.

Proof For $w \in \mathrm{int}(C) \cap T(f)$ let $\mathbf{C}_w$ be the set of f-cycles $K \subset C$ with $w \in \mathrm{int}(K)$ and $\mathrm{per}(K) = N$; so $\mathbf{C}_w \neq \varnothing$ for some $w \in \mathrm{int}(C) \cap T(f)$. If K, $K' \in \mathbf{C}_w$ then, as in the proof of Lemma 3.5, there exists $K'' \in \mathbf{C}_w$ with $K'' \subset K \cap K'$. (Note that if $K'' \subset K \cap K'$ is an f-cycle with $w \in \mathrm{int}(K'')$ then by the maximality of N we must have $\mathrm{per}(K'') = N$, and so $K'' \in \mathbf{C}_w$.) Now for $w \in \mathrm{int}(C) \cap T(f)$ with $\mathbf{C}_w \neq \varnothing$ let us put $K_w = \bigcap_{K\in\mathbf{C}_w} K$. Thus K_w is either an f-cycle with period N, or $w \in \mathrm{Per}(N,f)$ and $K_w = \{w,f(w),\ldots,f^{N-1}(w)\}$. Suppose the latter holds; then we can find $K \in \mathbf{C}_w$ with $K \cap T(f) \subset \{w,f(w),\ldots,f^{N-1}(w)\}$. But this is not possible because it would imply that either $(w,w+\varepsilon) \subset Z(f)$ or $(w-\varepsilon,w) \subset Z(f)$ for some $\varepsilon > 0$. Therefore K_w is an f-cycle. Suppose K_w is not minimal; then there exists an f-cycle $K \subset K_w$ with $K \neq K_w$. By construction $w \notin \mathrm{int}(K)$, and since $\mathrm{per}(K) \geq \mathrm{per}(K_w)$ we must have $\mathrm{per}(K) = N$; hence $K \in \mathbf{C}_v$ for some $v \in \mathrm{int}(C) \cap T(f)$ with $v \neq w$. This means that if K_w is not minimal then K_v is a proper subset of K_w for some other $v \in \mathrm{int}(C) \cap T(f)$. But $\mathrm{int}(C) \cap T(f)$ is finite, and thus K_u is minimal for some $u \in \mathrm{int}(C) \cap T(f)$. □

Lemma 3.7 If K is a minimal f-cycle with $K \cap Z(f) = \varnothing$ then K is topologically transitive.

Proof Let F be closed subset of K with $f(F) \subset F$, and suppose that $U = \text{int}(F) \neq \emptyset$. Let U_0 be a maximal connected component of U. For each $m \geq 1$ we have $f^m(U_0) \subset F$ and $f^m(U_0)$ is an interval; there thus exists a unique component U_m of U such that $f^m(U_0) \subset \overline{U_m}$. We cannot have $U_n \cap U_m = \emptyset$ for all $n > m \geq 0$, because this, exactly as in Lemma 3.1, would imply that $U_0 \cap Z(f) \neq \emptyset$. Therefore there exist $n > m \geq 0$ with $U_m = U_n$, and by Lemma 3.2 $K' = \bigcup_{j=m}^{n-1} \overline{U_j}$ is then an f-cycle. Since K is minimal we must have $K' = K$. But $K' \subset F$, and so $F = K$. Hence K is topologically transitive. □

Theorem 3.4 now follows immediately from Lemmas 3.5, 3.6 and 3.7. □

We now turn to the proof of Theorem 2.5, and show that each topologically transitive f-cycle is either exact or semi-exact. Let C be a topologically transitive f-cycle with period m and B be one of the m components of C; let g be the restriction of f^m to B. Then $g \in \mathbf{M}(B)$ and it is easy to see that g is topologically transitive (i.e. B is a topologically transitive g-cycle). Thus it is enough to show that if $f \in \mathbf{M}(I)$ is topologically transitive then f is either exact or semi-exact.

Let us say that $f \in \mathbf{M}(I)$ **splits** if there exists $c \in (a,b)$ such that $f([a,c]) = [c,b]$ and $f([c,b]) = [a,c]$, (and note then that $f(c) = c$). Suppose f splits, and let g be the restriction of f^2 to $[a,c]$; then $g \in \mathbf{M}([a,c])$ does not split, because $g(c) = c$. Now if f is topologically transitive then g is also topologically transitive, and if g is exact then f is semi-exact. The problem of showing that each topologically transitive cycle is either exact or semi-exact is therefore reduced to proving the following result:

Theorem 3.8 Let $f \in \mathbf{M}(I)$ be topologically transitive and suppose that f does not split. Then f is exact.

Proof We proceed via a number of lemmas.

Lemma 3.9 Let $f \in M(I)$ and $B_0,\dots, B_{p-1}$ be non-trivial closed intervals with $f(B_{k-1}) \subset B_k$ for $k = 1,\dots, p-1$ and $f(B_{p-1}) \subset B_0$. Let $Q = \bigcup_{k=0}^{p-1} B_k$ (and so by Lemma 3.2 Q is an f-cycle). If $\mathrm{int}(B_0),\dots, \mathrm{int}(B_{p-1})$ are disjoint then Q has period either p or $p/2$.

Proof If $B_0,\dots, B_{p-1}$ are disjoint then clearly Q has period p , and so we can assume that this is not the case. It then follows that there exists $0 < k < p$ such that $B_0 \cap B_j = \emptyset$ for $j = 1,\dots, k-1$ and $B_0 \cap B_k \neq \emptyset$. By induction we thus have $B_{nk} \cap B_{(n+1)k} \neq \emptyset$ for for all $n \geq 0$ (where for $n \geq p$ we define B_n so that $B_n = B_m$ whenever $n \equiv m \pmod p$). But $\mathrm{int}(B_{nk}) \cap \mathrm{int}(B_{(n+1)k}) = \emptyset$ for all $n \geq 0$ and $B_{kp} = B_0$; hence the only possibility is that $2k = p$. Q then has period $p/2$. □

Lemma 3.10 If $f \in M(I)$ is topologically transitive and does not split then $\bigcup_{n\geq 0} f^n(J) \supset (a,b)$ for each non-empty open interval $J \subset I$.

Proof Let $J \subset I$ be a non-empty open interval. Then exactly the same argument as at the beginning of the section (and using the fact that $Z(f) = \emptyset$) gives us an f-cycle $C = \bigcup_{k=0}^{q-1} \overline{V_k}$, where $V_0,\dots, V_{q-1}$ are components of $\mathrm{int}(\bigcup_{n\geq 0} f^n(J))$ with $f(\overline{V_{k-1}}) \subset \overline{V_k}$ for $k = 1,\dots, q-1$ and $f(\overline{V_{q-1}}) \subset \overline{V_0}$. Let $\ell = \min\{ k \geq 1 : f(\overline{V_k}) \subset \overline{V_0} \}$, then clearly $V_0,\dots, V_{\ell-1}$ are disjoint and $C = \bigcup_{k=0}^{\ell-1} \overline{V_k}$. But f is topologically transitive and so we must have $C = I$, and therefore by Lemma 3.9 ℓ is either 1 or 2 . If $\ell = 2$ then there exists $c \in (a,b)$ with $f([a,c]) \subset [c,b]$ and $f([c,b]) \subset [a,c]$, and since f is topologically transitive we would then have $f([a,c]) = [c,b]$ and $f([c,b]) = [a,c]$. This is not possible, because we are assuming f does not split, and hence $\ell = 1$. We thus have $\bigcup_{n\geq 0} f^n(J) \supset V_0 \supset (a,b)$. □

Lemma 3.11 Let $f \in M(I)$ be topologically transitive; then there exist $u , v \in T(f^2)$ with $f^2(u) = a$ and $f^2(v) = b$.

Proof Note that for any $g \in M(I)$ we have $\{ x \in I : g(x) = b \} \subset \{a,b\} \cup T(g)$.

Suppose that $f^2(a) = b$; then $f(f(a)) = b$ and so $f(a) \in \{b\} \cup T(f)$; ($f(a) = a$ is clearly not possible). If $f(a) = b$ then also $f(b) = b$, and hence $f(v) = a$ for some $v \in T(f)$ (since f is onto); thus $v \in T(f^2)$ and $f^2(v) = b$. If $f(a) = w \in T(f)$ then we can find $v \in (a,b)$ with $f(v) = w$; again we have $v \in T(f^2)$ and $f^2(v) = b$. Suppose now that $f^2(v) \neq b$ for all $v \in T(f^2)$; then by the above we must have $f^2(a) \neq b$, and so $f^2(b) = b$ (since f^2 is onto). Choose $d < b$ with $d \geq \max\{ f^2(z) : z \in \{a\} \cup T(f^2) \}$ such that f^2 is increasing on $[d,b]$. If $f^2(d) \geq d$ then $(d,b) \subset Z(f)$, which by Proposition 2.2 (3) is not possible; hence $f^2(d) < d$. But then we have $f^2([a,d]) \subset [a,d]$, and thus $f([a,d]) = [c,b]$ for some $c < d$ (because f is topologically transitive and $f(d) \neq d$); this is also not possible since it implies that $f([c,d]) \subset [c,d]$. Therefore $f^2(v) = b$ for some $v \in T(f^2)$, and in the same way $f^2(u) = a$ for some $u \in T(f^2)$. □

Let us assume for the rest of the proof that $f \in \mathbf{M}(I)$ is topologically transitive and that f does not split. Lemma 3.11 shows in particular that $f^2((a,b)) = I$, and thus by Lemma 3.10 we have $\bigcup_{n\geq 0} f^n(J) = I$ for each non-empty open interval $J \subset I$.

Let $z \in \mathrm{Fix}(f)$; then either f^2 is increasing in a neighbourhood of z or $z \in T(f^2)$, and so without loss of generality we can assume that f^2 is increasing on $[z,z+\varepsilon]$ for some $\varepsilon > 0$.

Lemma 3.12 $f^m([z,z+\varepsilon]) = I$ for some $m \geq 0$.

Proof Since $Z(f) = \emptyset$ we must have $f^2(x) > x$ for all $x \in (z,z+\varepsilon]$. Thus $f^2([z,z+\varepsilon]) \supset [z,z+\varepsilon]$ and this means that $\{f^{2n}([z,z+\varepsilon])\}_{n\geq 0}$ is an increasing sequence of closed intervals. Let $L = \overline{\bigcup_{n\geq 0} f^{2n}([z,z+\varepsilon])}$; then L is a closed interval with $f^2(L) \subset L$. But $z \in L \cap f(L)$, and therefore $L \cup f(L)$ is a non-trivial closed interval with $f(L \cup f(L)) \subset L \cup f(L)$. Therefore $L \cup f(L) = I$, because f is topologically transitive. Now consider the closed interval $J = L \cap f(L)$; we have $z \in J$ and $f(J) \subset J$, and hence we have either $J = I$ or $J = \{z\}$. Since f does not split the latter is not possible, and thus $J = I$, i.e. $L = I$. This gives us that $\bigcup_{n\geq 0} f^{2n}([z,z+\varepsilon]) \supset (a,b)$, and hence by Lemma 3.11 $f^m([z,z+\varepsilon]) = I$ for some $m \geq 0$. □

Now let $J \subset I$ be a non-empty open interval. Since $\bigcup_{n\geq 0} f^n(J) = I$ there exists $k \geq 0$ with $z \in f^k(J)$, and then $z \in f^n(J)$ for all $n \geq k$, because $z \in Fix(f)$. But $f^k(J)$ contains a non-empty open interval, and hence we also have $\bigcup_{n\geq k} f^n(J) = I$. There thus exists $j \geq k$ with $z+\varepsilon \in f^j(J)$. Therefore $f^j(J)$ is an interval containing z and $z+\varepsilon$, and so $f^j(J) \supset [z,z+\varepsilon]$. Hence by Lemma 3.12 $f^{j+m}(J) = I$ for some $m \geq 0$. This shows that f is exact, which completes the proof of Theorem 3.8. □

4. SINKS AND HOMTERVALS

In this section we analyse the set $Z(f)$. Recall that for $f \in M(I)$ we defined

$$Z(f) = \{ x \in (a,b) : \text{there exits } \varepsilon > 0 \text{ such that } f^n \text{ is monotone on } (x-\varepsilon,x+\varepsilon) \text{ for all } n \geq 0 \} ;$$

thus $Z(f)$ is open, and by Lemma 2.1 $f(Z(f)) \subset Z(f)$. Let

$$Z_*(f) = \{ x \in I : f^n(x) \in Z(f) \text{ for some } n \geq 0 \} ;$$

then it is clear that $Z_*(f)$ is open, $Z(f) \subset Z_*(f)$, and also $f(Z_*(f)) \subset Z_*(f)$ and $f(I-Z_*(f)) \subset I-Z_*(f)$.

Proposition 4.1 Let $f \in M(I)$; then each point in $Z_*(f)-Z(f)$ is an isolated point of $I-Z(f)$, and so in particular $Z_*(f)-Z(f)$ is countable.

Proof Let $x \in Z_*(f)-Z(f)$ and put $m = \min\{ n \geq 0 : f^n(x) \in Z(f) \}$; hence $m \geq 1$ and $x \in \{a,b\} \cup T(f^m)$. There thus exists an open neighbourhood U of x such that $f^m(U) \subset Z(f)$ and $U \cap (\{a,b\} \cup T(f^m)) = \{x\}$, and it follows that $U-\{x\} \subset Z(f)$; i.e. x is an isolated point of $I-Z(f)$. □

We will in fact analyse the set $Z_*(f)$ rather than $Z(f)$. Note that if B is an open subset of I then by Proposition 4.1 we have $Z(f) \cap B \neq \emptyset$ if and only if $Z_*(f) \cap B \neq \emptyset$; the same is also true when B is a non-trivial closed interval. We can thus replace $Z(f)$ by $Z_*(f)$ in Proposition 2.2 and in the definition of an f-register-shift.

Let $f \in M(I)$; we call a non-empty open interval $J \subset (a,b)$ a **sink** of f if there exists $m \geq 1$ such that f^m is monotone on J and $f^m(J) \subset J$. (If J is a sink of f then by (2.1) it follows that f^n is monotone on J for all $n \geq 0$.) A non-empty open interval $J \subset (a,b)$ is called a **homterval** of f if for each $n \geq 0$ we have f^n is monotone on J and $f^n(J)$ is not contained in any sink. Let

$$\mathrm{Sink}(f) = \{ x \in I : f^n(x) \in J \text{ for some sink } J \text{ and some } n \geq 0 \} ,$$

$$\mathrm{Homt}(f) = \{ x \in I : f^n(x) \in L \text{ for some homterval } L \text{ and some } n \geq 0 \} .$$

$Sink(f)$ and $Homt(f)$ are clearly both open.

Proposition 4.2 $Z_*(f) = Sink(f) \cup Homt(f)$ for each $f \in M(I)$.

Proof If J is either a sink or homterval of f then $J \subset Z(f)$, and thus $Sink(f) \cup Homt(f) \subset Z_*(f)$. Conversely, let U be a connected component of $Z(f)$; then $U \subset (a,b)$ is an open interval and f^n is monotone on U for each $n \geq 0$. Therefore U is either a homterval or $f^n(U) \subset J$ for some sink J and some $n \geq 0$, and this implies that $Z(f) \subset Sink(f) \cup Homt(f)$. Hence we also have $Z_*(f) \subset Sink(f) \cup Homt(f)$. □

There is a strong connection between sinks and "attracting" periodic points. To see this, we need a couple of simple facts about the periodic points of mappings from $M(I)$. Let $f \in M(I)$ and $x \in Per(m,f)$; we denote by $[x]$ the periodic orbit containing x , i.e. $[x] = \{x, f(x), \ldots, f^{m-1}(x)\}$. We consider $[x]$ as a subset of I , and thus if x , $y \in Per(m,f)$ then $[x] = [y]$ if and only if $y = f^k(x)$ for some $0 \leq k < m$. Now let $\alpha([x],f)$ denote the set of points in I which are attracted to the orbit $[x]$, i.e.

$$\alpha([x],f) = \{\, y \in I : \lim_{n\to\infty} f^{mn}(y) = f^k(x) \text{ for some } 0 \leq k < m \,\} .$$

Also let $\delta([x],f) = \{\, y \in I : f^n(y) = x$ for some $n \geq 0 \,\}$; thus $\delta([x],f)$ consists of those points in I which eventually hit the orbit $[x]$. Clearly $\delta([x],f) \subset \alpha([x],f)$, and $\delta([x],f)$ is countable (because $f^{-1}(\{z\})$ is finite for each $z \in I$).

Proposition 4.3 Let $f \in M(I)$ and $x \in Per(m,f)$. Then either

(4.1) $\alpha([x],f) = \delta([x],f)$, (in which case $\alpha([x],f)$ is countable), or

(4.2) there exists a non-trivial interval J with $x \in J \subset \alpha([x],f)$ such that $int(J)$ is a sink of f .

Moreover, if (4.2) holds then $\alpha([x],f) - \delta([x],f)$ is a non-empty open subset of $Sink(f)$.

Proof Let $g = f^{2m}$; then, since $x \in Fix(f^m)$, either g is increasing in a neighbourhood of x or $x \in T(g)$. Let us assume first that $x \in (a,b)$, and that g is increasing on $[u,v]$ for some $u < x < v$. Now (4.2) will hold if $g(y) < y$ for all $y \in (x,t)$ for some $t \in (x,v]$; and in the same way (4.2) will also hold if $g(y) > y$ for all $y \in (s,x)$ for some $s \in [u,x)$. Thus if (4.2) does not hold then for each $t \in (x,v]$ there exists $y \in (x,t)$ with $g(y) \geq y$, and for each $s \in [u,x)$ there exists $y \in (s,x)$ with $g(y) \leq y$. But this implies that if (4.2) does not hold then $\alpha([x],g) = \delta([x],g)$, and hence that (4.1) holds. Furthermore, the above argument shows also that $\alpha([x],f) - \delta([x],f)$ is an open subset of $\text{Sink}(f)$. The other cases (i.e. with $x \in \{a,b\}$ or $x \in T(g)$) are similar, and are left to be dealt with by the reader. □

Let $f \in M(I)$ and $x \in \text{Per}(m,f)$; we say that x is **attracting** if (4.2) holds. If x is an attracting periodic point of f then, in particular, $x \in \bar{J}$ for some sink J of f , and by Proposition 4.3 $\alpha([x],f) - \delta([x],f)$ is a non-empty open subset of $\text{Sink}(f)$.

We next consider the converse of the above, and look at whether attracting periodic points occur when we have a sink. Let J be a sink of $f \in M(I)$, and let m be the smallest positive integer such that $f^m(J) \subset J$ and f^m is increasing on J . (If k is the smallest positive integer such that $f^k(J) \subset J$ then m is either k or $2k$, depending on whether f^k is increasing or decreasing on J .) Then the sequence $\{f^{mn}(x)\}_{n\geq 0}$ is monotone for each $x \in J$, and thus for each $x \in J$ there exists $z \in \text{Per}(f)$ with $x \in \alpha([z],f)$. In general we cannot say much about these periodic points; however, if $\text{Fix}(f^m)$ is finite then it is easy to see that there exist finitely many attracting periodic points $x_1,\dots, x_p$ of f such that $J - \bigcup_{j=1}^{p} \alpha([x_j],f)$ is finite . This leads to the following result:

Proposition 4.4 Let $f \in M(I)$ and suppose $\text{Fix}(f^m)$ is finite for each $m \geq 1$. Let M be the set of attracting periodic points of f , (and note that M is countable); put $\alpha(f) = \bigcup_{x\in M} \alpha([x],f)$. Then $\text{Sink}(f) - \alpha(f)$ and $\alpha(f) - \text{Sink}(f)$ are both countable.

Proof Proposition 4.3 gives us immediately that $\alpha(f) - \text{Sink}(f)$ is countable. On the other hand, by the remark above we have that $J - \alpha(f)$ is finite for each sink

J of f, and thus $\mathrm{Sink}(f) - \alpha(f)$ is also countable. □

Remark: The assumption in Proposition 4.4 (that $\mathrm{Fix}(f^m)$ be finite for each $m \geq 1$) will clearly be satisfied if f is the restriction to I of an analytic function defined in some (complex) neighbourhood of I. Most mappings which have been used in applications are of this type (for example, polynomial mappings).

We now start looking at some of the general properties of sinks and homtervals.

Proposition 4.5 Let $f \in M(I)$.

(1) If J_1, J_2 are sinks (resp. homtervals) of f with $J_1 \cap J_2 \neq \emptyset$ then $J_1 \cup J_2$ is also a sink (resp. homterval) of f.

(2) If J is a sink (resp. homterval) of f then $f^n(J)$ is also a sink (resp. homterval) for each $n \geq 0$.

(3) If J is a sink and L a homterval of f then $J \cap L = \emptyset$.

(4) Each sink is contained in a maximal sink; two maximal sinks are either equal or disjoint.

(5) Each homterval is contained in a maximal homterval; two maximal homtervals are either equal or disjoint.

(6) If L is a homterval of f then the sets $\{f^n(L)\}_{n\geq 0}$ are disjoint.

Proof (1) is clear, and (2) follows since if $J \subset (a,b)$ is an open interval with f^n monotone on J then $f^n(J) \subset (a,b)$ and $f^n(J)$ is open. For the other parts we need a lemma:

Lemma 4.6 Suppose $K \subset (a,b)$ is a non-empty open interval such that f^n is monotone on K for all $n \geq 0$. If $K \cap f^m(K) \neq \emptyset$ for some $m \geq 1$ then K is contained in a sink of f.

Proof Let $J = \bigcup_{j\geq 0} f^{jm}(K)$, so $K \subset J$. It is easily checked that J is a sink of f. □

(3): Suppose $J \cap L \neq \emptyset$; then applying Lemma 4.6 to $K = J \cup L$ would give us that

K , and thus also L , is contained in a sink of f . Since L is a homterval this is not possible.

(4): Let J be a sink of f and let K be the largest open interval with $J \subset K \subset (a,b)$ such that f^n is monotone on K for all $n \geq 0$. Lemma 4.6 and the maximality of K show that K is the maximal sink containing J . If J_1, J_2 are maximal sinks with $J_1 \neq J_2$ then (1) immediately gives us that $J_1 \cap J_2 = \emptyset$.

(5): Let L be a homterval of f and let K be the largest open interval with $L \subset K \subset (a,b)$ such that f^n is monotone on K for all $n \geq 0$. Then K is the required maximal homterval. The rest follows as in (4).

(6): Let $0 \leq j < k$; then applying Lemma 4.6 with $K = f^j(L)$ and $m = k-j$ shows that $f^j(L) \cap f^k(L) = \emptyset$. □

Proposition 4.5 gives us that $\mathrm{Sink}(f) \cap \mathrm{Homt}(f) = \emptyset$ and that $f(\mathrm{Sink}(f)) \subset \mathrm{Sink}(f)$ and $f(\mathrm{Homt}(f)) \subset \mathrm{Homt}(f)$.

Let $f \in M(I)$ with topologically transitive f-cycles $C_1, \ldots, C_r$ and f-register-shifts $R_1, \ldots, R_\ell$. Then by Propositions 2.3, 4.1, 4.2 and 4.5 and Theorem 2.4 we have that the sets

$$A(C_1,f), \ldots, A(C_r,f),\ A(R_1,f), \ldots, A(R_\ell,f),\ \mathrm{Sink}(f) \text{ and } \mathrm{Homt}(f)$$

are disjoint and their union is a dense subset of I .

We now analyse the set Sink(f) . Let $J = (c,d)$ be a sink of $f \in M(I)$, and let m be the smallest positive integer such that $f^m(J) \subset J$ and f^m is increasing on J . Let [u,v] be the largest interval containing J on which f^m is increasing. We call J a **trap** of f if $u < c < d < v$ and $f^m(z) < z$ for all $z \in [u,c)$, $f^m(z) > z$ for all $z \in (d,v]$. (Note then that $c , d \in \mathrm{Fix}(f^m)$.) A trap is clearly a maximal sink. (See the picture on the following page.)

We say that a maximal sink J of f is **central** if one of the end-points of $\bar{J}$ is a turning point of f (i.e. if $T(f) \cap \bar{J} \neq \emptyset$). Finally, we say that a maximal sink J of f is an **end** of f if $\bar{J} \cap \{a,b\} \neq \emptyset$ and $f^n(J)$ is not contained in a central sink for each $n \geq 0$.

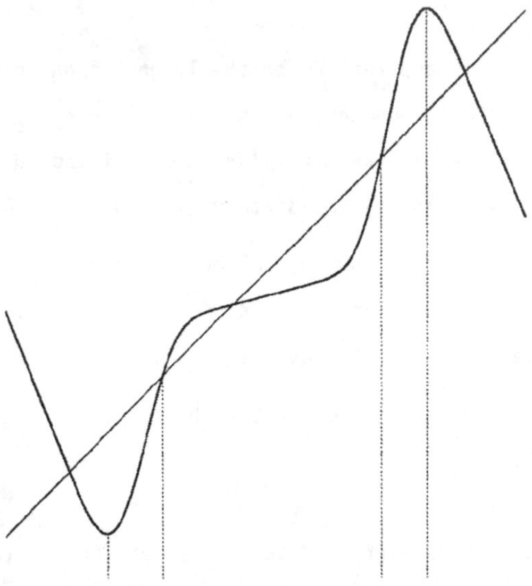

Lemma 4.7 Let $f \in M(I)$.

(1) If J is a maximal sink of f then J is either a trap, an end, or $f^n(J)$ is contained in a central sink for some $n \geq 0$.

(2) If J is a trap of f then so is $f(J)$.

(3) If J is an end of f then $f(J)$ is contained in an end.

Proof (1): Let $J = (c,d)$ be a maximal sink of f and m be the smallest positive integer such that $f^m(J) \subset J$ and f^m is increasing on J . Let $[u,v]$ be the largest interval containing J on which f^m is increasing. By the maximality of J we have $f^m(z) < z$ for all $z \in [u,c)$ and $f^m(z) > z$ for all $z \in (d,v]$. Thus if J is not a trap then either $u = c$ or $d = v$. Suppose that $u = c$. If $a < u$ then $u \in T(f^m)$ and hence by (2.1) there exists $0 \leq n < m$ with $f^n(u) \in T(f)$. Therefore $T(f) \cap f^n(J) \neq \emptyset$ and so $f^n(J)$ is contained in a central sink. If $a = u$ then either J is an end or $f^n(J)$ is contained in a central sink for some $n \geq 0$. The case when $d = v$ is dealt with in the same way.

(2): Let $J = (c,d)$ be a trap of f , and let m and $[u,v]$ be as in (1). f is monotone on $[u,v]$; without loss of generality we can assume that f is increasing on this interval. Thus $f(J) = (f(c),f(d))$; let $[s,t]$ be the largest interval containing $f(J)$ on which f^m is increasing. Since c , $d \in \mathrm{Fix}(f^m)$ we can find $c' \in (u,c)$ and $d' \in (d,v)$ so that $f^m((c',d')) \subset (u,v)$; then f^{2m} is increasing on (c',d') , which implies that f^m is increasing on $(f(c'),f(d'))$, and this gives us that $s < f(c) < f(d) < t$. Let K be the maximal sink containing $f(J)$; if $f(J)$ is not a trap then we would have that $f(J)$ is a proper subset of K (because $s < f(c) < f(d) < t$). But this would imply that $J = f^m(J)$ is a proper subset of $f^{m-1}(K)$, and thus that J is not maximal. This is not possible, and so $f(J)$ is a trap.

(3): Let K be the maximal sink containg $f(J)$. By (2) K cannot be a trap (since for some $m \geq 1$ we have $f^{m-1}(K) \cap J \neq \emptyset$). Also, if $f^n(K)$ were contained in a central sink then so would $f^{n+1}(J)$ be, and this is not possible because J is an end. Therefore by (1) K is an end. □

If J is an end of $f \in M(I)$ then by Lemma 4.7 (3) we have either $f(J) \subset J$ or $f^2(J) \subset J$, i.e. J has period either 1 or 2 . Let u (resp. v) be the smallest (resp. largest) element of $T(f)$. It is straightforward to show that an end exists if and only if one of the following holds:

(4.3) f is increasing on $[a,u]$, $f(u) > u$ and $f(c) = c$ for some $c \in (a,u)$. (Then (a,c) is an end with period 1 , where c is the largest fixed point of f in (a,u) .)

(4.4) f is increasing on $[v,b]$, $f(v) < v$ and $f(d) = d$ for some $d \in (v,b)$. (Then (d,b) is an end with period 1 , where d is the smallest fixed point of f in (v,b) .)

(4.5) f is decreasing on $[a,u]$ and $[v,b]$ and (4.3) and (4.4) hold when f is replaced by f^2 and u (resp. v) by the smallest (resp. largest) element of $T(f^2)$. (Here we have two ends, each having period 2 .)

Note that if $|T(f)|$ is odd then there is at most one end.

For $f \in M(I)$ let

$$\mathrm{Trap}(f) = \{ x \in I : f^n(x) \in J \text{ for some trap } J \text{ and some } n \geq 0 \},$$

$$\mathrm{Cent}(f) = \{\, x \in I : f^n(x) \in J \text{ for some central sink } J \text{ and some } n \geq 0 \,\},$$

$$\mathrm{End}(f) = \{\, x \in I : f^n(x) \in J \text{ for some end } J \text{ and some } n \geq 0 \,\}.$$

Proposition 4.8 Let $f \in \mathbf{M}(I)$; then $\mathrm{Trap}(f)$, $\mathrm{Cent}(f)$ and $\mathrm{End}(f)$ are disjoint open sets with $\mathrm{Sink}(f) = \mathrm{Trap}(f) \cup \mathrm{Cent}(f) \cup \mathrm{End}(f)$.

Proof The sets $\mathrm{Trap}(f)$, $\mathrm{Cent}(f)$ and $\mathrm{End}(f)$ are clearly open; that they are disjoint follows from Lemma 4.7 and the fact that maximal sinks are either equal or disjoint. The last part follows from Lemma 4.7 (1) and Proposition 4.5 (4). □

Let $f \in \mathbf{M}(I)$, J be a central sink of f , and let m be the smallest positive integer such that $f^m(J) \subset J$; for $k = 0,\ldots, m-1$ let J_k be the maximal sink containing $f^k(J)$. Then it is easy to see that the intervals $\overline{J_0},\ldots, \overline{J_{m-1}}$ are disjoint, and thus $\bigcup_{k=0}^{m-1} \overline{J_k}$ is an f-cycle with period m . We call an f-cycle C **central** if it can be obtained in this way. Note that there are at most $|T(f)|$ central f-cycles. Similarly, we say that an f-cycle C is an **end** f-cycle if either $C = \overline{J}$, where J is an end with $f(J) \subset J$, or $C = \overline{J_1} \cup \overline{J_2}$, where J_1 and J_2 are ends with $f(J_1) \subset J_2$ and $f(J_2) \subset J_1$.

Lemma 4.9 (1) Two central f-cycles are either equal or disjoint.

(2) Let $G' = A(C_1',f) \cup \cdots \cup A(C_s',f)$, where $C_1',\ldots, C_s'$ are the central f-cycles; then $\mathrm{int}(\mathrm{Cent}(f) - G')$ and $\mathrm{int}(G' - \mathrm{Cent}(f))$ are both empty.

(3) Two end f-cycles are either equal or disjoint.

(4) Let $G'' = A(C_1'',f) \cup \cdots \cup A(C_e'',f)$, where $C_1'',\ldots, C_e''$ (with $0 \leq e \leq 2$) are the end f-cycles; then $\mathrm{int}(\mathrm{End}(f) - G'')$ and $\mathrm{int}(G'' - \mathrm{End}(f))$ are both empty.

Proof Easy exercise. □

Let $f \in \mathbf{M}(I)$ and $w \in T(f)$; we call w **central** if $w \in C$ for some central f-cycle C . Put $T_c(f) = \{\, w \in T(f) : w \text{ is central} \,\}$. We can now reformulate Theorem 2.4.

Theorem 4.10 Let $f \in M(I)$, let $C_1, \ldots, C_r$ be the topologically transitive f-cycles, $C'_1, \ldots, C'_s$ the central f-cycles, $C''_1, \ldots, C''_e$ the end f-cycles and $R_1, \ldots, R_\ell$ the f-register-shifts. Then $\ell + r + |T_c(f)| \leq |T(f)|$, the sets $A(C_1,f), \ldots, A(C_r,f)$, $A(C'_1,f), \ldots, A(C'_s,f)$, $A(C''_1,f), \ldots, A(C''_e,f)$, $A(R_1,f), \ldots, A(R_\ell,f)$, $Trap(f)$ and $Homt(f)$ are disjoint and their union is a dense subset of I .

Proof This follows immediately from Theorem 2.4, Proposition 2.2 and the results of this section. □

5. EXAMPLES OF REGISTER-SHIFTS

We now give a couple of simple examples to demonstrate that register-shifts actually occur. The mappings we consider are not smooth, but they have the advantage of being defined explicitly, and they are very easy to analyse.

It is also possible to obtain examples of register-shifts by showing that they must occur within a suitably chosen family of mappings. For instance, it can be shown that there exists a p_μ-register-shift for infinitely many values of $\mu \in (0,4]$, where $p_\mu : [0,1] \to [0,1]$ is given by $p_\mu(x) = \mu x(1-x)$. This approach can be found in Feigenbaum (1978), (1979), Misiurewicz (1981), Collet, Eckmann and Lanford (1980) and Jonker and Rand (1981).

Our first example of a mapping having a register-shift follows. It is similar to an example in Milnor and Thurston (1977). Let $I = [0,1]$, and for each $n \geq 0$ let $a_n = \frac{1}{2}(1-3^{-n})$ and put $f(a_n) = \frac{4}{5}(1-6^{-n})$. Now define f to be linear on each interval $[a_n, a_{n+1}]$, $n \geq 0$ (and so f has slope 2^{-n+1} on $[a_n, a_{n+1}]$). This defines f as a strictly increasing, continuous function on $[0,\frac{1}{2})$. Put $f(\frac{1}{2}) = \frac{4}{5}$ and let $f(x) = f(1-x)$ for $x \in (\frac{1}{2},1]$. Then $f \in \mathbf{M}([0,1])$, $f(0) = f(1) = 0$ and $\frac{1}{2}$ is the single turning point of f. We will show that there exists an f-register-shift R such that $[0,1] - A(R,f)$ is countable.

Lemma 5.1 $f^2([\frac{1}{3},\frac{2}{3}]) \subset [\frac{1}{3},\frac{2}{3}]$, and if g is the restriction of f^2 to $[\frac{1}{3},\frac{2}{3}]$ then $g(x) = \frac{1}{3}(2 - f(2-3x))$ for all $x \in [\frac{1}{3},\frac{2}{3}]$, i.e. $g = \psi \circ f \circ \psi^{-1}$, where $\psi : [0,1] \to [\frac{1}{3},\frac{2}{3}]$ is the linear change of variables given by $\psi(t) = \frac{1}{3}(2-t)$. (This means that g is just f turned upside down and scaled down by a factor of 3. See the picture on the next page.)

Proof We have $f([\frac{1}{3},\frac{2}{3}]) = [\frac{2}{3},\frac{4}{5}]$ and $f([\frac{2}{3},\frac{4}{5}]) = [\frac{2}{5},\frac{2}{3}]$; hence $f^2([\frac{1}{3},\frac{2}{3}]) \subset [\frac{1}{3},\frac{2}{3}]$. For each $n \geq 0$ we have $f(a_{n+1}) = \frac{2}{3} + \frac{1}{6}f(a_n)$ and thus, since f is linear on $[a_n, a_{n+1}]$, $f(\frac{1}{2}-t) = \frac{2}{3} + \frac{1}{6}f(\frac{1}{2}-3t)$ for all $t \in [0,\frac{1}{6}]$. Therefore $f(x) = \frac{2}{3} + \frac{1}{6}f(3x-1) = \frac{2}{3} + \frac{1}{6}f(2-3x)$ for all $x \in [\frac{1}{3},\frac{1}{2}]$, and so also for all $x \in [\frac{1}{3},\frac{2}{3}]$. But if $x \in [\frac{1}{3},\frac{2}{3}]$ then $f(x) \in [\frac{2}{3},1]$, and hence $f^2(x) = 2 - 2f(x)$; this implies that $f^2(x) = \frac{1}{3}(2 - f(2-3x))$ for all $x \in [\frac{1}{3},\frac{2}{3}]$. □

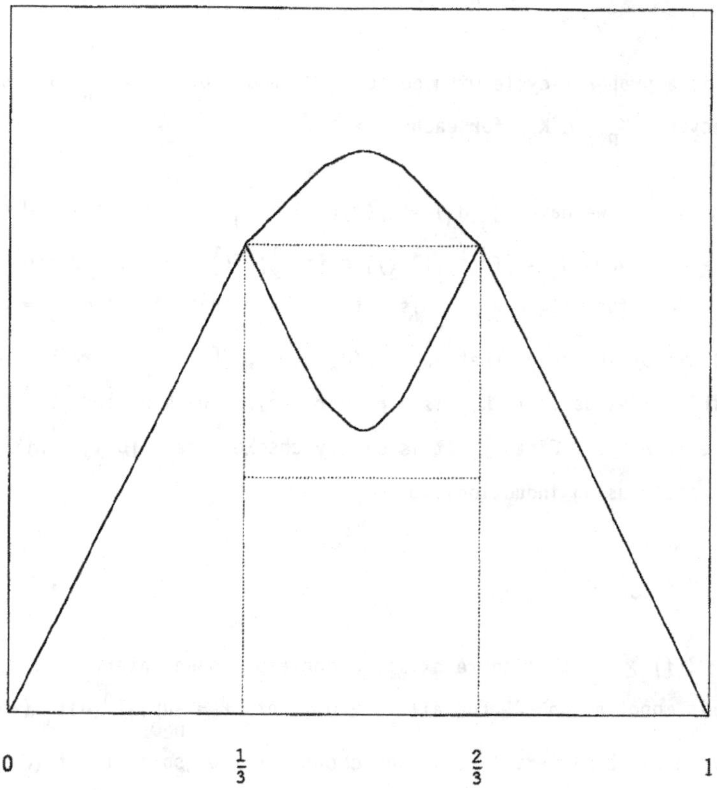

Lemma 5.2 For each $n \geq 1$ we have $f^{2^n}([a_n, 1-a_n]) \subset [a_n, 1-a_n]$, and if g_n is the restriction of f^{2^n} to $[a_n, 1-a_n]$ then $g_n = \psi_n \circ f \circ \psi_n^{-1}$, where $\psi_n : [0,1] \to [a_n, 1-a_n]$ is the linear change of variables given by

$$\psi_n(t) = \begin{cases} (1-a_n) - (1-2a_n)t & \text{if } n \text{ is odd,} \\ a_n + (1-2a_n)t & \text{if } n \text{ is even.} \end{cases}$$

Proof This follows by induction from Lemma 5.1. (Note that $[\frac{1}{3}, \frac{2}{3}] = [a_1, 1-a_1]$ and $\psi = \psi_1$.) □

Let $J = [f^2(\frac{1}{2}), f(\frac{1}{2})] = [\frac{2}{5}, \frac{4}{5}]$; then $f(J) = J$, and so J is a proper f-cycle with period 1 . For each $n \geq 1$ let $J_n = \psi_n(J)$ and $K_n = \bigcup_{j=0}^{2^n-1} f^j(J_n)$.

Lemma 5.3 K_n is a proper f-cycle with period 2^n and $[0,1] - A(K_n,f)$ is countable; moreover, $K_{n+1} \subset K_n$ for each $n \geq 1$.

Proof For each $n \geq 1$ we have $g_n(J_n) = g_n(\psi_n(J)) = \psi_n(f(J)) = \psi_n(J) = J_n$, and hence $f^{2^n}(J_n) = J_n$. Now $J_1 = [f^2(\frac{1}{2}),f^4(\frac{1}{2})] \subset [f^2(\frac{1}{2}),f(\frac{1}{2})] = J$, and thus by induction $J_{n+1} \subset J_n$ for all $n \geq 1$. Also it is clear that $f(J_1) \cap J_1 = \emptyset$, and so it again follows by induction that J_n , $f(J_n)$,..., $f^{2^n-1}(J_n)$ are disjoint for each $n \geq 1$. This gives us that K_n is a proper f-cycle with period 2^n and $K_{n+1} \subset K_n$ for all $n \geq 1$. Finally, it is easily checked that $[0,1] - A(K_n,f)$ is countable (once again using induction). □

Lemma 5.4 $Z(f) = \emptyset$.

Proof Suppose $Z(f) \neq \emptyset$; then there exists a non-empty open interval $J \subset (0,1)$ such that f^n is monotone on J for all $n \geq 0$. Let $\ell = \sup_{n\geq 0} |f^n(J)|$ (where $|L|$ denotes the length of the interval L), and choose $k \geq 0$ so that $|f^k(J)| > \ell/2$. Then for some $m \geq 0$ we must have either $a_m \in f^k(J)$ or $1-a_m \in f^k(J)$: If this were not the case then there exists $p \geq 0$ such that either $f^k(J) \subset (a_p,a_{p+1})$ or $f^k(J) \subset (1-a_{p+1},1-a_p)$. Now f^{2^p} is linear with slope 2 (resp. with slope -2) on $[a_p,a_{p+1}]$ (resp. on $[1-a_{p+1},1-a_p]$), and hence $f^{k+2^p}(J)$ would be twice as long as $f^k(J)$, which is not possible. However, if $\{a_m,1-a_m\} \cap f^k(J) \neq \emptyset$ then there exists $\varepsilon > 0$ such that f^n is monotone on $[0,\varepsilon)$ for all $n \geq 0$ (since the restriction of f^{2^m} to $[a_m,1-a_m]$ is just a "small copy" of f , and if f^n is monotone on $(1-\delta,1]$ for all $n \geq 0$ for some $\delta > 0$ then f^n is also monotone on $[0,\varepsilon)$ for all $n \geq 0$ for some $\varepsilon > 0$). But this cannot happen, because for each $\varepsilon > 0$ there exists $w \in (0,\varepsilon)$ and $n \geq 1$ such that $f^n(w) = \frac{1}{2}$. Thus we have $Z(f) = \emptyset$. □

Let $R = \bigcap_{n\geq 1} K_n$; then by Lemmas 5.3 and 5.4 R is an f-register- shift, and

$\{K_n\}_{n\geq 1}$ is a generator for R . Moreover, $[0,1] - A(R,f)$ is countable, because $[0,1] - A(R,f) = \bigcup_{n\geq 1} ([0,1] - A(K_n,f))$.

Remarks: (1) For $\mu \in (0,4]$ let $p_\mu : [0,1] \to [0,1]$ be given by $p_\mu(x) = \mu x(1-x)$. Then for a certain value of μ (about 3.569946) there also exists a p_μ-register-shift R_μ such that $R_\mu = \bigcap_{n\geq 1} K_n^\mu$, where K_n^μ is a proper p_μ-cycle with period 2^n , and such that $[0,1] - A(R_\mu,p_\mu)$ is countable (see the references at the beginning of the section). This mapping p_μ and the mapping f are in fact conjugate (see Proposition 12.10).

(2) The construction of the mapping f can be generalized somewhat: Fix $0 < \alpha < 1$, and for $n \geq 0$ let $b_n = \frac{1}{2}(1-\alpha^n)$ and $f_\alpha(b_n) = \frac{1}{2}(1+\alpha^2)^{-1}(1+\alpha)^2(1-\gamma^n)$, where $\gamma = \alpha(1+\alpha)^{-1}(1-\alpha)$. Now define f_α to be linear on $[b_n,b_{n+1}]$ for each $n \geq 0$, let $f_\alpha(x) = f_\alpha(1-x)$ for $x \in (\frac{1}{2},1]$, and put $f_\alpha(\frac{1}{2}) = \frac{1}{2}(1+\alpha^2)^{-1}(1+\alpha)^2$. In particular, the mapping f is obtained by taking $\alpha = \frac{1}{3}$. The same argument which was used for f now shows that for each $\alpha \in (0,1)$ there exists an f_α-register-shift R_α such that $[0,1] - A(R_\alpha,f_\alpha)$ is countable, and again we have $R_\alpha = \bigcap_{n\geq 1} K_n^\alpha$, where K_n^α is a proper f_α-cycle with period 2^n . However, it is easy to see that all the mappings f_α , $0 < \alpha < 1$, are conjugate.

Let $f \in M(I)$ and R be an f-register-shift; we say that R is **tame** if there exists an f-cycle K with $R \subset K$ such that $K - A(R,f)$ is countable. Our first example of a register-shift was thus tame. In Section 9 we show that the structure of a tame register-shift is somewhat special. To be more precise, let R be a tame f-register-shift and K be an f-cycle with $R \subset K$ such that $K - A(R,f)$ is countable. Then it follows from Theorem 9.7 that each point $x \in K - A(R,f)$ is eventually periodic (i.e. $f^m(x) \in Per(f)$ for some $m \geq 0$), and if $x \in K \cap Per(f)$ then x has period $2^j\,per(K)$ for some $j \geq 0$. Moreover, if $\{K_n\}_{n\geq 1}$ is a generator for R with $K_1 \subset K$ then $per(K_n) = 2^{q_n}\,per(K)$ for all $n \geq 1$ for some strictly increasing sequence $\{q_n\}_{n\geq 1}$ of non-negative integers.

We now modify our first example to obtain a non-tame register- shift. Again let $I = [0,1]$ and for $n \geq 0$ let $a_n = \frac{1}{2}(1-7^{-n})$ and $f(a_n) = \frac{24}{27}(1-28^{-n})$. Define f to be linear on each interval $[a_n,a_{n+1}]$, $n \geq 0$ (and so f has slope $2\cdot 4^{-n}$ on $[a_n,a_{n+1}]$). As before this defines f as a strictly increasing, continuous function on $[0,\frac{1}{2})$. Put $f(\frac{1}{2}) = \frac{24}{27}$ and for $x \in (\frac{1}{2},1]$ let $f(x) = f(1-x)$. Then $f \in M(I)$, $f(0) = f(1) = 0$ and $\frac{1}{2}$ is the single turning point of f .

Lemma 5.5 $f^3([\frac{3}{7},\frac{4}{7}]) \subset [\frac{3}{7},\frac{4}{7}]$, and if g is the restriction of f^3 to $[\frac{3}{7},\frac{4}{7}]$ then $g(x) = \frac{1}{7}(4 - f(4-7x))$, i.e. $g = \psi \circ f \circ \psi^{-1}$, where $\psi : [0,1] \to [\frac{3}{7},\frac{4}{7}]$ is the linear change of variables given by $\psi(t) = \frac{1}{7}(4-t)$. (This means that g is just f turned upside down and scaled down by a factor of 7 .)

Proof This is similar to the calculation in the proof of Lemma 5.1. Note that $f([\frac{3}{7},\frac{4}{7}]) = [\frac{6}{7},\frac{24}{27}]$, $f([\frac{6}{7},\frac{24}{27}]) = [\frac{6}{27},\frac{2}{7}]$ and $f([\frac{6}{27},\frac{2}{7}]) = [\frac{12}{27},\frac{4}{7}] \subset [\frac{3}{7},\frac{4}{7}]$; also f is linear with slope -2 on $[\frac{6}{7},\frac{24}{27}]$ and is linear with slope 2 on $[\frac{6}{27},\frac{2}{7}]$. □

Lemma 5.6 For each $n \geq 1$ we have $f^{3^n}([a_n,1-a_n]) \subset [a_n,1-a_n]$ and if g_n is the restriction of f^{3^n} to $[a_n,1-a_n]$ then $g_n = \psi_n \circ f \circ \psi_n^{-1}$, where $\psi_n : [0,1] \to [a_n,1-a_n]$ is the linear change of variables given by

$$\psi_n(t) = \begin{cases} (1-a_n) - (1-2a_n)t & \text{if } n \text{ is odd,} \\ a_n + (1-2a_n)t & \text{if } n \text{ is even.} \end{cases}$$

Proof This follows by induction from Lemma 5.5. □

Let $J = [f^2(\frac{1}{2}),f(\frac{1}{2})] = [\frac{6}{27},\frac{24}{27}]$; then $f(J) = J$, and so J is a proper f-cycle with period 1 . For each $n \geq 1$ let $J_n = \psi_n(J)$ and $K_n = \bigcup_{j=0}^{3^n-1} f^j(J_n)$.

Lemma 5.7 K_n is a proper f-cycle with period 3^n ; moreover, $K_{n+1} \subset K_n$ for each $n \geq 1$.

Proof Similar to the proof of Lemma 5.3. □

Lemma 5.8 $Z(f) = \emptyset$.

Proof Almost exactly the same as the proof of Lemma 5.4. □

Let $R = \bigcap_{n \geq 1} K_n$; then by Lemmas 5.7 and 5.8 R is an f-register- shift, and $\{K_n\}_{n \geq 1}$ is a generator for R . Note that by Theorem 2.4 $A(R,f)$ is dense in I . We now show that R is not tame.

Lemma 5.9 $J - A(K_1,f)$ is uncountable, (where again $J = [\frac{6}{27},\frac{24}{27}]$).

Proof If $x \in A(K_1,f)$ then there exists $m \geq 0$ such that $f^m(x) \in [\frac{3}{7},\frac{4}{7}]$, and hence $f^{m+1}(x) \notin [\frac{1}{3},\frac{2}{3}]$ and $f^{m+2}(x) \notin [\frac{1}{3},\frac{2}{3}]$. Therefore $J - A(K_1,f) \supset B$, where $B = \{ x \in [\frac{1}{3},\frac{2}{3}] : f^{2n}(x) \in [\frac{1}{3},\frac{2}{3}]$ for all $n \geq 1 \}$. Consider f^2 on $[\frac{1}{3},\frac{2}{3}]$. f^2 is decreasing on $[\frac{1}{3},\frac{1}{2}]$ and increasing on $[\frac{1}{2},\frac{2}{3}]$ and $f^2(\frac{1}{3}) = f^2(\frac{2}{3}) = \frac{2}{3}$. Moreover, f^2 is linear with slope -4 on $[\frac{1}{3},\frac{3}{7}]$ and is linear with slope 4 on $[\frac{4}{7},\frac{2}{3}]$. Thus in fact f^2 maps each of the two intervals $[\frac{1}{3},\frac{5}{12}]$ and $[\frac{7}{12},\frac{2}{3}]$ linearly onto $[\frac{1}{3},\frac{2}{3}]$, and $f^2((\frac{5}{12},\frac{7}{12})) \cap [\frac{1}{3},\frac{2}{3}] = \emptyset$.

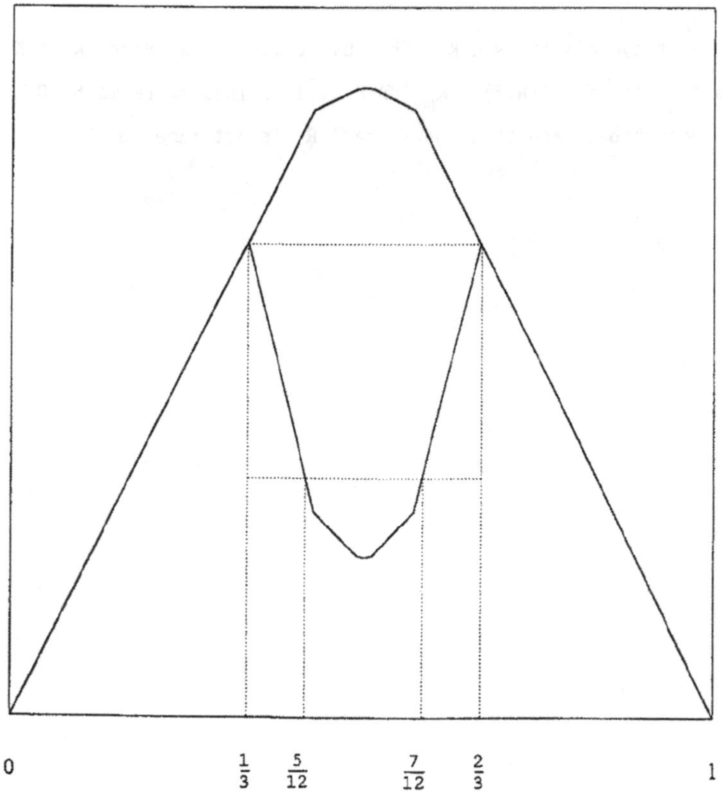

For $m \geq 1$ let $B_m = \{ x \in [\frac{1}{3},\frac{2}{3}] : f^{2n}(x) \in [\frac{1}{3},\frac{2}{3}]$ for $1 \leq n \leq m \}$, hence $(B_m)_{m\geq 1}$ is a decreasing sequence of closed sets with $B = \bigcap_{m\geq 1} B_m$. We have $B_1 = [\frac{1}{3},\frac{5}{12}] \cup [\frac{7}{12},\frac{2}{3}]$, and it is easily checked that in general B_m consists of 2^m closed intervals, each having length $\frac{1}{3}\cdot 4^{-m}$, and that B_{m+1} is obtained by deleting the "middle half" of each of the intervals in B_m . Thus B is a Cantor set, and so in particular B is uncountable. Therefore $J - A(K_1,f)$ is also uncountable. □

Lemma 5.10 $K_m - A(K_{m+1},f)$ is uncountable for each $m \geq 1$.

Proof By Lemma 5.9 $J - A(K_1,f)$ is uncountable, and hence $\psi_m(J) - A(\psi_m(K_1),g_m)$ is uncountable. But $\psi_m(J)$ is a component of K_m and $\psi_m(K_1) = K_{m+1} \cap \psi_m(J)$. Thus $(\psi_m(J) - A(\psi_m(K_1),g_m)) - (K_m - A(K_{m+1},f))$ is countable, and so $K_m - A(K_{m+1},f)$ is uncountable. □

Now let K be any f-cycle with $R \subset K$. Then by Lemma 2.15 we have $K_m \subset K$ for some $m \geq 1$, and hence $K - A(R,f) \supset K_m - A(K_{m+1},f)$. Thus by Lemma 5.10 $K - A(R,f)$ is uncountable, and this shows that R is not tame.

6. A PROOF OF PARRY'S THEOREM (THEOREM 2.6)

Recall that $g \in M(I)$ is said to be **uniformly piecewise linear with slope** β ($\beta > 0$) if on each of its laps g is linear with slope β or $-\beta$. In this section we give a proof of the following result:

Theorem 6.1 (Parry (1966)) If $f \in M(I)$ is topologically transitive then f is conjugate to a uniformly piecewise linear mapping.

Note that by Theorem 2.5 Theorem 6.1 is really the same as Theorem 2.6. The proof of Theorem 6.1 which we give here is taken, with a few minor modifications, from Milnor and Thurston (1977).

Let $V(I) = \{ \psi \in C(I) : \psi$ is increasing and onto $\}$ (where by increasing we mean only that $\psi(x) \geq \psi(y)$ whenever $x \geq y$). If $\psi \in V(I)$ and $g \in M(I)$ then we say that (ψ,g) is a **reduction** (or semi-conjugacy) of $f \in M(I)$ if $\psi \circ f = g \circ \psi$. In Sections 7 and 8 we will analyse the reductions of a mapping $f \in M(I)$ in some detail; the reason for introducing this concept here is because of the following simple fact:

Lemma 6.2 Let (ψ,g) be a reduction of $f \in M(I)$. If f is topologically transitive then ψ must be a homeomorphism, and so in particular f and g are conjugate.

Proof Suppose ψ is not a homeomorphism; then there exist $x, y \in I$ with $x < y$ and $\psi(x) = \psi(y)$. Put $z = \psi(x)$, and for $n \geq 0$ let $J_n = \psi^{-1}(\{g^n(z)\})$. We thus have $f(J_n) \subset J_{n+1}$ for each $n \geq 0$ and $[x,y] \subset J_0$; hence $\{J_n\}_{n\geq 0}$ is a sequence of non-trivial closed intervals. Let $F = \overline{\bigcup_{n\geq 1} J_n}$; then F is a closed subset of I with $f(F) \subset F$ and $\mathrm{int}(F) \neq \emptyset$, and therefore $F = I$, because f is topologically transitive. This means that $J_0 \cap J_m \neq \emptyset$ for some $m \geq 1$, and thus $J_0 = J_m$ (since for each j , $k \geq 0$ either $J_j = J_k$ or $J_j \cap J_k = \emptyset$). Let $p = \min\{ m \geq 1 : J_0 = J_m \}$; then $J_0, \ldots, J_{p-1}$ are disjoint, and $F' = \bigcup_{k=0}^{p-1} J_k$ is a closed subset of I with $f(F') \subset F'$ and $\mathrm{int}(F') \neq \emptyset$. It again follows that $F' = I$, and hence we must have $p = 1$, i.e. $J_0 = I$. However, this is clearly not possible, because ψ is onto, and so ψ is a homeomorphism. □

Lemma 6.2 reduces the proof of Theorem 6.1 to showing that if $f \in M(I)$ is topologically transitive then there exists a reduction (ψ, g) of f with g uniformly piecewise linear.

For $f \in M(I)$ let $\ell(f)$ denote the number of laps of f, so $\ell(f) = |T(f)| + 1$; also let $h(f) = \inf_{n \geq 1} \frac{1}{n} \log \ell(f^n)$, and thus $h(f) \geq 0$.

Lemma 6.3 $h(f) = \lim_{n \to \infty} \frac{1}{n} \log \ell(f^n)$ for each $f \in M(I)$.

Proof If $f, g \in M(I)$ then $\ell(f \circ g) \leq \ell(f)\ell(g)$, since each lap of g contains at most $\ell(f)$ laps of $f \circ g$, and so in particular $\ell(f^{m+n}) \leq \ell(f^m)\ell(f^n)$ for all $m, n \geq 1$. Put $a_n = \log \ell(f^n)$; then $a_{m+n} \leq a_m + a_n$ for all $m, n \geq 1$, and hence $\lim_{n \to \infty} \frac{1}{n} a_n = \inf_{n \geq 1} \frac{1}{n} a_n$. (A proof of this very well-known fact is given in Lemma 13.4.) □

Remark: If $|T(f)| > 0$ then we also have $h(f) = \lim_{n \to \infty} \frac{1}{n} \log |T(f^n)|$ (since $\ell(f^n) = |T(f^n)| + 1$). In Misiurewicz and Szlenk (1980) it is shown that $h(f)$ is the topological entropy of f.

Lemma 6.4 Let $f \in M(I)$, and suppose there exists a topologically transitive f-cycle; then $h(f) > 0$. In particular, $h(f) > 0$ if f is topologically transitive.

Proof Let C be a topologically transitive f-cycle; put $m = \mathrm{per}(C)$, let B be one of the m components of C, and let g be the restriction of f^m to B. By Theorem 2.5 $g \in M(I)$ is either exact or semi-exact, and hence we can find $p \geq 1$ and $D, E \subset B$ with $D \cap E = \varnothing$ such that $g^p(D) = g^p(E) = B$. Thus g^p is at least 2 to 1 (i.e. for each $x \in B$ we have $|(g^p)^{-1}(\{x\})| \geq 2$), and so g^{pn} is at least 2^n to 1 for each $n \geq 1$. It follows that $\ell(g^{pn}) \geq 2^n$ for all $n \geq 1$, and hence $\ell(f^{mpn}) \geq \ell(g^{pn}) \geq 2^n$ for all $n \geq 1$. Therefore by Lemma 6.3

$$h(f) = \lim_{n\to\infty} \frac{1}{mpn} \log \ell(f^{mpn}) \geq \frac{1}{mp} \log 2 > 0 . \qquad \square$$

By Lemmas 6.2 and 6.4 we now have that Theorem 6.1 is a corollary of the following result of Milnor and Thurston:

Theorem 6.5 Let $f \in \mathbf{M}(I)$ with $h(f) > 0$; then there exists a reduction (ψ,g) of f such that g is uniformly piecewise linear with slope β , where $\beta = \exp(h(f))$.

Proof The proof which we give is essentially that to be found in Milnor and Thurston (1977). Fix $f \in \mathbf{M}(I)$ with $h(f) > 0$ and put $r = \exp(-h(f))$; thus $r = 1/\beta$ and $0 < r < 1$. Note that by Lemma 6.3 we have $\beta = \lim_{n\to\infty} \ell(f^n)^{1/n}$, and hence r is the radius of convergence of the power series $\sum_{n\geq 0} \ell(f^n)t^n$; in particular the series $L(t) = \sum_{n\geq 0} \ell(f^n)t^n$ converges for all $t \in (0,r)$.

Lemma 6.6 $\lim_{t\uparrow r} L(t) = \infty$.

Proof We have $\ell(f^n) \geq [\exp(h(f))]^n = \beta^n$ for each $n \geq 0$, and hence $L(t) \geq \sum_{n\geq 0} (\beta t)^n = r(r-t)^{-1}$ for all $t \in (0,r)$. $\square$

Let $\mathbf{J}$ denote the set of non-trivial closed intervals $J \subset I$; for $J \in \mathbf{J}$ and $n \geq 0$ we denote by $\ell(f^n|J)$ the number of laps of f^n which intersect the interior of J (and so in fact $\ell(f^n|J) = |T(f^n) \cap \mathrm{int}(J)| + 1$). Then $\ell(f^n|J) \leq \ell(f^n|I) = \ell(f^n)$, and thus in particular the series $L(J,t) = \sum_{n\geq 0} \ell(f^n|J)t^n$ converges for all $t \in (0,r)$. Now since $L(I,t) = L(t) \neq 0$ we can define $\Lambda(J,t) = L(J,t)/L(I,t)$ for each $t \in (0,r)$, and we have $0 \leq \Lambda(J,t) \leq 1$, because $L(J,t) \leq L(I,t)$.

Lemma 6.7 Let J , $K \in \mathbf{J}$ and suppose J and K intersect in a single point. Then

$$\lim_{t\uparrow r} | \Lambda(J \cup K,t) - \Lambda(J,t) - \Lambda(K,t) | = 0 .$$

Proof For each $n \geq 0$ we have

$$\ell(f^n|J) + \ell(f^n|K) - 1 \leq \ell(f^n|J \cup K) \leq \ell(f^n|J) + \ell(f^n|K) ,$$

and thus

$$\begin{aligned} | \Lambda(J \cup K,t) - \Lambda(J,t) - \Lambda(K,t) | &= L(I,t)^{-1} \, | L(J \cup K,t) - L(J,t) - L(K,t) | \\ &\leq L(I,t)^{-1} \sum_{n\geq 0} t^n = [L(t)(1-t)]^{-1} . \end{aligned}$$

But by Lemma 6.6 we have $\lim_{t\uparrow r} [L(t)(1-t)]^{-1} = 0$ (since $r < 1$). □

Lemma 6.8 Let $J \in J$ be such that f is monotone on J . Then

$$\lim_{t\uparrow r} | r\Lambda(f(J),t) - \Lambda(J,t) | = 0 .$$

Proof Since f is monotone on f we have $\ell(f^{n+1}|J) = \ell(f^n|f(J))$ for each $n \geq 0$, and thus

$$\begin{aligned} L(J,f) &= \sum_{n\geq 0} \ell(f^n|J)t^n = 1 + \sum_{n\geq 0} \ell(f^{n+1}|J)t^{n+1} \\ &= 1 + t \sum_{n\geq 0} \ell(f^n|f(J))t^n = 1 + tL(f(J),t) . \end{aligned}$$

Hence

$$\begin{aligned} | r\Lambda(f(J),t) - \Lambda(J,t) | &= L(I,t)^{-1} \, | rL(f(J),t) - L(J,t) | \\ &\leq L(I,t)^{-1} \left(|rL(f(J),t)-tL(f(J),t)| + |tL(f(J),t)-L(J,t)|\right) \\ &\leq |r-t| + L(I,t)^{-1} = |r-t| + L(t)^{-1} , \end{aligned}$$

and by Lemma 6.6 $\lim_{t\uparrow r} (|r-t| + L(t)^{-1}) = 0$. □

Lemma 6.9 Let $J \in J$ and $m \geq 1$. If f^m is monotone on J then $\limsup_{t\uparrow r} \Lambda(J,t) \leq r^m$.

Proof For $k = 0,\ldots, m-1$ we have f is monotone on $f^k(J)$, and thus by Lemma 6.8

$$\lim_{t\uparrow r} | r^{k+1}\Lambda(f^{k+1}(J),t) - r^k\Lambda(f^k(J),t) |$$

$$= \lim_{t\uparrow r} r^k | r\Lambda(f(f^k(J)),t) - \Lambda(f^k(J),t) | = 0 .$$

Hence $\lim_{t\uparrow r} | r^m\Lambda(f^m(J),t) - \Lambda(J,t) | = 0$, and so

$$\limsup_{t\uparrow r} \Lambda(J,t) = \limsup_{t\uparrow r} r^m\Lambda(f^m(J),t) \le r^m ,$$

since $\Lambda(f^m(J),t) \le 1$ for all $t \in (0,r)$. □

Lemma 6.10 There exists a sequence $\{t_n\}_{n\ge1}$ from $(0,r)$ with $\lim_{n\to\infty} t_n = r$ such that $\{\Lambda(J,t_n)\}_{n\ge1}$ converges for all $J \in \mathbf{J}$.

Proof Let I_0 be a countable dense subset of I with $\{a,b\} \subset I_0$ and $T(f^n) \subset I_0$ for each $n \ge 1$; let $\mathbf{J_0}$ be the set of intervals $J = [c,d] \in \mathbf{J}$ such that $c , d \in I_0$, thus $\mathbf{J_0}$ is countable. Now if $J \in \mathbf{J}$ and $\{s_n\}_{n\ge1}$ is any sequence from $(0,r)$ then the sequence $\{\Lambda(J,s_n)\}_{n\ge1}$ is bounded, and so there exists a subsequence $\{n_k\}_{k\ge1}$ such that $\{\Lambda(J,s_{n_k})\}_{k\ge1}$ converges. Therefore, since $\mathbf{J_0}$ is countable, we can find (using the standard diagonal argument) a sequence $\{t_n\}_{n\ge1}$ from $(0,r)$ with $\lim_{n\to\infty} t_n = r$ such that $\{\Lambda(J,t_n)\}_{n\ge1}$ converges for every $J \in \mathbf{J_0}$. We show then that this sequence converges for all $J \in \mathbf{J}$: First consider $J = [c,d] \in \mathbf{J}$ with $c \in I_0$ and $d \notin I_0$. Let $\varepsilon > 0$ and choose $m \ge 1$ so that $r^m < \varepsilon$; since $d \notin T(f^m) \cup \{a,b\}$ there exist $u , v \in I_0$ such that $c < u < d < v$ and f^m is monotone on $[u,v]$. Then $[c,u]$ and $[c,v]$ are both in $\mathbf{J_0}$, and $\Lambda([c,u],t) \le \Lambda(J,t) \le \Lambda([c,v],t)$ for all $t \in (0,r)$; hence by Lemmas 6.7 and 6.9

$$\limsup_{n\to\infty} \Lambda(J,t_n) - \liminf_{n\to\infty} \Lambda(J,t_n) \le \lim_{n\to\infty} \Lambda([c,v],t_n) - \lim_{n\to\infty} \Lambda([c,u],t_n)$$

$$= \lim_{n\to\infty} \Lambda([u,v],t_n) \le r^m < \varepsilon .$$

Therefore, since $\varepsilon > 0$ was arbitrary, the sequence $\{\Lambda(J,t_n)\}_{n\ge1}$ converges. The same argument also gives that this sequence converges when $J = [c,d]$ with $c \notin I_0$ and $d \in I_0$. Finally, if $J = [c,d]$ and $c \notin I_0$, $d \notin I_0$ then choose $u \in I_0$ with $c < u < d$; we then have that $\{\Lambda([c,u],t_n)\}_{n\ge1}$ and $\{\Lambda([u,d],t_n)\}_{n\ge1}$ both converge, and so by Lemma 6.7 $\{\Lambda(J,t_n)\}_{n\ge1}$ converges. □

Now fix a sequence $\{t_n\}_{n\geq 1}$ as in Lemma 6.10, and for each $J \in \mathbf{J}$ define $\Lambda(J) = \lim_{n\to\infty} \Lambda(J,t_n)$. By Lemmas 6.7, 6.8 and 6.9 we then have:

(6.1) If J , $K \in \mathbf{J}$ and $J \cap K$ consists of a single point then $\Lambda(J \cup K) = \Lambda(J) + \Lambda(K)$.

(6.2) If $J \in \mathbf{J}$ and f is monotone on J then $r\Lambda(f(J)) = \Lambda(J)$.

(6.3) If $J \in \mathbf{J}$, $m \geq 1$ and f^m is monotone on J then $\Lambda(J) \leq r^m$.

We also define a mapping $\pi : I \to [0,1]$ by letting $\pi(a) = 0$ and $\pi(x) = \Lambda([a,x])$ for $x \in (a,b]$.

Lemma 6.11 The mapping $\pi : I \to [0,1]$ is continuous, increasing and onto.

Proof It is clear that π is increasing, and if it is continuous then it is onto, because $\pi(a) = 0$ and $\pi(b) = \Lambda(I) = 1$. Let $x \in I$ and $\varepsilon > 0$; choose $m \geq 1$ so that $r^m < \varepsilon$. Then there exists $\delta > 0$ such that $\{ w \in T(f^m) : |w-x| < \delta \} \subset \{x\}$. Now if we put $U = I \cap (x-\delta,x+\delta)$ then U is a neighbourhood of x in I , and it follows from (6.1) and (6.3) that $|\pi(y)-\pi(x)| < \varepsilon$ for all $y \in U$, since if $y > x$ (resp. $y < x$) then f^m is monotone on $[x,y]$ (resp. on $[y,x]$). This shows that π is continuous. □

Lemma 6.12 There exists a unique mapping $\alpha : [0,1] \to [0,1]$ such that $\pi\circ f = \alpha\circ\pi$.

Proof If $\alpha : [0,1] \to [0,1]$ is a mapping with $\pi\circ f = \alpha\circ\pi$ then $\alpha(z) = \pi(f(x))$ whenever $x \in I$ is such that $\pi(x) = z$. Conversely, we can use this relation to define a mapping α with $\pi\circ f = \alpha\circ\pi$, provided we have $\pi(f(x)) = \pi(f(y))$ whenever x , $y \in I$ are such that $\pi(x) = \pi(y)$. Let x , $y \in I$ with $x < y$ and $\pi(x) = \pi(y)$, and consider u , v with $x \leq u < v \leq y$ so that f is monotone on $[u,v]$. Then $\pi(u) = \pi(v)$ and hence by (6.1) and (6.2) we have

$$r\Lambda(f([u,v])) = \Lambda([u,v]) = \Lambda([a,v]) - \Lambda([a,u]) = \pi(v) - \pi(u) = 0 ,$$

i.e. $\Lambda(f([u,v])) = 0$; thus $\pi(f(u)) = \pi(f(v))$, because $f(u)$ and $f(v)$ are the end-points of the interval $f([u,v])$. But we can write

$$[x,y] = [u_0,u_n] = [u_0,u_1] \cup [u_1,u_2] \cup \cdots \cup [u_{n-1},u_n]$$

with f monotone on each of the intervals $[u_j,u_{j+1}]$, $0 \le j \le n-1$, and therefore

$$\pi(f(x)) = \pi(f(u_0)) = \pi(f(u_1)) = \cdots = \pi(f(u_n)) = \pi(f(y)) .$$

This shows that α exists. The uniqueness of α follows immediately from the fact that π is onto. □

Let $\alpha : [0,1] \to [0,1]$ be the unique mapping with $\pi\circ f = \alpha\circ\pi$.

Lemma 6.13 α is uniformly piecewise linear with slope β .

Proof Let $[c,d]$ be a lap of f on which f is increasing, and let $z \in (\pi(c),\pi(d)]$. Then there exists $x \in (c,d]$ with $\pi(x) = z$, and hence by (6.1) and (6.2) we have

$$\begin{aligned} \alpha(z) - \alpha(\pi(c)) &= \alpha(\pi(x)) - \alpha(\pi(c)) = \pi(f(x)) - \pi(f(c)) \\ &= \Lambda([a,f(x)]) - \Lambda([a,f(c)]) = \Lambda([f(c),f(x)]) = \Lambda(f([c,x])) \\ &= \beta\Lambda([c,x]) = \beta(\Lambda([a,x]) - \Lambda([a,c])) = \beta(\pi(x) - \pi(c)) = \beta(z - \pi(c)) . \end{aligned}$$

This shows that α is linear with slope β on $[\pi(c),\pi(d)]$. If $[u,v]$ is a lap of f on which f is decreasing then a similar calculation shows that α is linear on $[\pi(u),\pi(v)]$ with slope $-\beta$. Therefore α is uniformly piecewise linear with slope β . □

Now let $\gamma : [0,1] \to I$ be the linear rescaling given by $\gamma(t) = a + (b-a)t$, and define ψ , $g : I \to I$ by $\psi = \gamma\circ\pi$ and $g = \gamma\circ\alpha\circ\gamma^{-1}$. Then $\psi \in V(I)$ and $g \in \mathbf{M}(I)$ is uniformly piecewise linear with slope β (because α is); moreover, we have $\psi\circ f = \gamma\circ\pi\circ f = \gamma\circ\alpha\circ\pi = \gamma\circ\alpha\circ\gamma^{-1}\circ\gamma\circ\pi = g\circ\psi$, i.e. (ψ,g) is a reduction of f . This completes the proof of Theorem 6.5. □

Remark: In Milnor and Thurston (1977) it is shown that for each $J \in \mathbf{J}$ there exists a meromorphic function $L_1(J,\cdot) : D = \{ z \in \mathbb{C} : |z| < 1 \} \to \mathbb{C} \cup \{\infty\}$ with $L_1(J,t) = L(J,t)$ for all $t \in (0,r)$. There thus also exists a meromorphic function $\Lambda_1(J,\cdot) : D \to \mathbb{C} \cup \{\infty\}$ with $\Lambda_1(J,t) = \Lambda(J,t)$ for all $t \in (0,r)$ ($\Lambda_1(J,\cdot) = L_1(J,\cdot)/L_1(I,\cdot)$). Now since $0 \le \Lambda(J,t) \le 1$ for all $t \in (0,r)$, we must have $0 \le \Lambda_1(J,r) \le 1$ and $\lim_{t\uparrow r} \Lambda(J,t) = \Lambda_1(J,r)$. In particular, we have $\Lambda(J) = \Lambda_1(J,r)$. The construction in Lemma 6.10 is therefore not really necessary.

We end this section by giving a result of Misiurewicz and Szlenk (Misiurewicz and Szlenk (1980)), which provides an alternative method of calculating $h(f)$ for a mapping $f \in \mathbf{M}(I)$. This will show in particular that if $g \in \mathbf{M}(I)$ is uniformly piecewise linear with slope β , $\beta \ge 1$, then $h(g) = \log \beta$.

For $f \in \mathbf{C}(I)$ let

$$\mathrm{Var}(f) = \sup \{ \sum_{k=0}^{n-1} |f(x_{k+1})-f(x_k)| \; : \; a = a_0 < x_1 < \cdots < x_n = b \} .$$

If $f \in \mathbf{M}(I)$ with $T(f) = \{d_1,\ldots,d_N\}$, where $a = d_0 < d_1 < \cdots < d_N < d_{N+1} = b$, then clearly

$$\mathrm{Var}(f) = \sum_{k=0}^{N} |f(d_{k+1})-f(d_k)| . \tag{6.4}$$

In particular, if $g \in \mathbf{M}(I)$ is uniformly piecewise linear with slope β ($\beta > 0$) then $\mathrm{Var}(g) = (b-a)\beta$.

Theorem 6.14 (Misiurewicz and Szlenk (1980)) Let $f \in \mathbf{M}(I)$; then $h(f) > 0$ if and only if $\limsup_{n\to\infty} \frac{1}{n} \log \mathrm{Var}(f^n) > 0$. Moreover, if $h(f) > 0$ then

$$h(f) = \lim_{n\to\infty} \frac{1}{n} \log \mathrm{Var}(f^n) .$$

Remark: Let $g \in \mathbf{M}(I)$ be uniformly piecewise linear with slope β , $\beta > 0$. Then $\mathrm{Var}(g^n) = (b-a)\beta^n$ for each $n \ge 1$, since g^n is uniformly piecewise linear with slope β^n , and thus $\lim_{n\to\infty} \frac{1}{n} \log \mathrm{Var}(g^n) = \log \beta$. Hence by Theorem 6.14 we have that $h(g) = \log \beta$, provided $\beta \ge 1$.

Proof By (6.4) we immediately have that $\mathrm{Var}(f^n) \le (b-a)\ell(f^n)$ for each $n \ge 1$, and hence $\limsup_{n\to\infty} \frac{1}{n} \log \mathrm{Var}(f^n) \le h(f)$. We can thus assume that $h(f) > 0$, and need to prove that $\liminf_{n\to\infty} \frac{1}{n} \log \mathrm{Var}(f^n) \ge h(f)$.

Let $g \in \mathbf{M}(I)$, $n \ge 1$, $\varepsilon > 0$ and $x_1,\ldots, x_m \in I$; $\{x_1,\ldots,x_m\}$ is called

(n,ε)-**separated** with respect to g if $\max_{0\le k<n} |g^k(x_i)-g^k(x_j)| > \varepsilon$ for all $1 \le i < j \le m$.

Lemma 6.15 Let $\{x_1,\dots,x_m\}$ be (n,ε)-separated with repect to $g \in M(I)$; then $(m-1)\varepsilon \le \sum_{k=0}^{n-1} \mathrm{Var}(g^k)$.

Proof We can assume that $x_1 < x_2 < \cdots < x_m$. Now for each i , $1 \le i \le m-1$, there exists $0 \le k \le n-1$ such that $|g^k(x_{i+1})-g^k(x_i)| > \varepsilon$, and from this the result immediately follows. □

Let $\{x_1,\dots,x_m\}$ be (n,ε)-separated with respect to $g \in M(I)$; $\{x_1,\dots,x_m\}$ is called **maximally** (n,ε)-separated if $\{x_1,\dots,x_m,x\}$ is not (n,ε)-separated for each $x \in I$. By Lemma 6.15 there exist maximally (n,ε)-separated sets.

Lemma 6.16 Let $g \in M(I)$ with laps $I_1,\dots, I_p$, and let $\varepsilon = \frac{1}{2}\min\{ |I_j| : 1 \le j \le p \}$; let $n \ge 1$ and $\{x_1,\dots,x_s\}$ be maximally (n,ε)-separated with respect to g . Then $\ell(g^n) \le 3^n s$.

Proof Let $J_1,\dots, J_q$ be the laps of g^n . If $1 \le i \le q$ and $0 \le k < n$ then there exists a unique $\alpha(i,k) \in \{1,\dots,p\}$ such that $g^k(J_i) \subset I_{\alpha(i,k)}$. We thus have a mapping $\alpha : \{1,\dots,q\} \to \{1,\dots,p\}^n$ defined by $\alpha(i) = (\alpha(i,0),\dots,\alpha(i,n-1))$, and this mapping α is injective. Now for $j = 1,\dots, s$ let $B_j = \{ x \in I : |g^k(x)-g^k(x_j)| \le \varepsilon$ for each $k = 0,\dots, n-1 \}$. Then $I = \bigcup_{j=1}^{s} B_j$, since $\{x_1,\dots,x_s\}$ is maximally (n,ε)-separated with respect to g . Hence, letting c_i denote the middle point of the interval J_i , we can choose $\sigma(i) \in \{1,\dots,s\}$ such that $c_i \in B_{\sigma(i)}$. This gives us a mapping $\sigma : \{1,\dots,q\} \to \{1,\dots,s\}$. Suppose $\sigma(i) = \sigma(j)$; then $|g^k(c_i)-g^k(c_j)| \le \varepsilon$ for $k = 0,\dots, n-1$ and this implies that $g^k(c_i)$ and $g^k(c_j)$ are either in the same or are in adjacent laps of g for each $k = 0,\dots, n-1$, i.e. $|\alpha(i,k)-\alpha(j,k)| \le 1$ for each $k = 0,\dots, n-1$ (assuming that the laps of g have been numbered in the usual way). Therefore $|\{ 1 \le j \le q : \sigma(j) = \sigma(i) \}| \le 3^n$ for each $i \in \{1,\dots,q\}$, and so we have $\ell(g^n) = q \le 3^n s$. □

Lemma 6.17 Let $g \in \mathbf{M}(I)$; then $\ell(g^n) \le 3^n M \sum_{k=0}^{n-1} \mathrm{Var}(g^k)$ for all $n \ge 1$ for some $M \ge 0$.

Proof Let $\varepsilon > 0$ be as in Lemma 6.16; let $n \ge 1$ and $\{x_1,\dots,x_s\}$ be maximally (n,ε)-separated with repect to g . By Lemma 6.16 we have $\ell(g^n) \le 3^n s$, and by Lemma 6.15 $s \le 1 + \frac{1}{\varepsilon} \sum_{k=0}^{n-1} \mathrm{Var}(g^k)$. Put $M = \frac{1}{\varepsilon} + (b-a)^{-1}$; then $1 + \frac{1}{\varepsilon} \sum_{k=0}^{n-1} \mathrm{Var}(g^k) \le M \sum_{k=0}^{n-1} \mathrm{Var}(g^k)$ for all $n \ge 1$, because $\sum_{k=0}^{n-1} \mathrm{Var}(g^k) \ge \mathrm{Var}(g^0) = b-a$, and hence $\ell(g^n) \le 3^n M \sum_{k=0}^{n-1} \mathrm{Var}(g^k)$ for all $n \ge 1$. □

Lemma 6.18 Let $g \in \mathbf{M}(I)$ be onto (i.e. $g(I) = I$); then $\mathrm{Var}(g^{n+1}) \ge \mathrm{Var}(g^n)$ for all $n \ge 0$.

Proof Fix $n \ge 0$ and let $a = d_0 < d_1 < \cdots < d_N < d_{N+1} = b$ be such that $\mathrm{Var}(g^n) = \sum_{k=0}^{N} |g^n(d_{k+1})-g^n(d_k)|$. Since g is onto there exist $a \le x_0 < x_1 < \cdots < x_{N+1} \le b$ such that $d_k = g(x_k)$ for $k = 0,\dots, N+1$, and hence

$$\mathrm{Var}(g^n) = \sum_{k=0}^{N} |g^{n+1}(x_{k+1})-g^{n+1}(x_k)| \le \mathrm{Var}(g^{n+1}) . \qquad \square$$

Proposition 6.19 Let $f \in \mathbf{M}(I)$ be onto; then $\liminf_{n\to\infty} \frac{1}{n} \log \mathrm{Var}(f^n) \ge h(f)$.

Proof The result is trivially true when $h(f) = 0$ (because $\mathrm{Var}(f^n) \ge b-a$ for all $n \ge 0$); hence we can assume that $h(f) > 0$. Let $0 < \varepsilon < \frac{1}{2}h(f)$, and choose $m \ge 1$ with $\frac{1}{m} \log 3 < \varepsilon$. By Lemma 6.17 there exists $M > 0$ such that $\ell(f^{mn}) \le 3^n M \sum_{k=0}^{n-1} \mathrm{Var}(f^{mk})$ for all $n \ge 1$, and since f^m is onto we then have by Lemma 6.18 that $\ell(f^{mn}) \le 3^n Mn\, \mathrm{Var}(f^{mn})$ for all $n \ge 1$. Let $p \ge 1$ be such that $\frac{1}{mn} \log(Mn) < \varepsilon$ and $\frac{n}{n+1} \ge 1-\varepsilon$ for all $n \ge p$. Thus if $n \ge p$ then

$$\begin{aligned} h(f) &\le \frac{1}{mn} \log \ell(f^{mn}) \le \frac{1}{m} \log 3 + \frac{1}{mn} \log (Mn) + \frac{1}{mn} \log \mathrm{Var}(f^{mn}) \\ &< 2\varepsilon + \frac{1}{mn} \log \mathrm{Var}(f^{mn}) . \end{aligned}$$

Now let $s \geq mp$; we can write $s = mn + r$ with $n \geq p$ and $0 \leq r < m$, and hence

$$\frac{1}{s} \log \operatorname{Var}(f^s) \geq \frac{mn}{s} \cdot \frac{1}{mn} \log \operatorname{Var}(f^{mn}) \geq \frac{mn}{s} (h(f) - 2\varepsilon)$$
$$\geq \frac{n}{n+1} (h(f) - 2\varepsilon) \geq (1-\varepsilon)(h(f) - 2\varepsilon) .$$

Therefore $\liminf_{n\to\infty} \frac{1}{n} \log \operatorname{Var}(f^n) \geq (1-\varepsilon)(h(f)-2\varepsilon)$ for all $\varepsilon \in (0, \frac{1}{2}h(f))$, and so $\liminf_{n\to\infty} \frac{1}{n} \log \operatorname{Var}(f^n) \geq h(f)$. □

Now let $f \in M(I)$ with $h(f) > 0$; by Theorem 6.5 there then exists a reduction (ψ, g) of f such that g is uniformly piecewise linear with slope $\beta = \exp(h(f))$. Let $J = \bigcap_{n\geq 0} f^n(I)$ and $J' = \psi(J)$; then, since $\{f^n(I)\}_{n\geq 0}$ is a decreasing sequence of compact sets, we have

$$J' = \psi(\bigcap_{n\geq 0} f^n(I)) = \bigcap_{n\geq 0} \psi(f^n(I)) = \bigcap_{n\geq 0} g^n(\psi(I)) = \bigcap_{n\geq 0} g^n(I) .$$

Thus J' is a non-trivial interval. ($\{g^n(I)\}_{n\geq 0}$ is a decreasing sequence of non-trivial closed intervals; if $J' = \{z\}$ for some $z \in I$ then $z \in \operatorname{Fix}(g)$, and this is not possible because $\beta > 1$.) Hence J is also a non-trivial interval, and $f(J) = J$. Let u be the restriction of f to J ; then $u \in M(J)$ is onto, and so by Proposition 6.19

$$\liminf_{n\to\infty} \frac{1}{n} \log \operatorname{Var}(f^n) \geq \liminf_{n\to\infty} \frac{1}{n} \log \operatorname{Var}(u^n) \geq h(u) .$$

Let v be the restriction of g to J' ; if K' is a lap of v^n then $\psi(K) \subset K'$ for some lap K of u^n , and this implies that $\ell(u^n) \geq \ell(v^n)$ for each $n \geq 0$; in particular we have $h(u) \geq h(v)$. But v^n is uniformly piecewise linear with slope β^n , and therefore each lap of v^n can have length at most $\beta^{-n}|J'|$. Thus $\ell(v^n) \geq \beta^n$ for each $n \geq 0$, and so $h(v) \geq \log \beta$. Hence we have

$$\liminf_{n\to\infty} \frac{1}{n} \log \operatorname{Var}(f^n) \geq h(u) \geq h(v) \geq \log \beta = \log(\exp(h(f))) = h(f) ,$$

and this completes the proof of Theorem 6.14. □

7. REDUCTIONS

We now start the second stage in the analysis of the iterates of a mapping $f \in \mathbf{M}(I)$. So far the main result has been Theorem 2.4, which gives us information about the asymptotic behaviour of the orbits $\{f^n(x)\}_{n\geq 0}$ for all points x lying in a residual subset $\Lambda(f)$ of I . The set $I-\Lambda(f)$, being the complement of a residual set, is topologically "small"; however, the behaviour of f on this set can strongly influence the global complexity of the iterates of f . In order to study this we introduce a construction which allows us to examine the action of f on various subsets of I .

As in Section 6 let $V(I) = \{ \psi \in C(I) : \psi$ is increasing and onto $\}$, where by increasing we mean only that $\psi(x) \geq \psi(y)$ whenever $x \geq y$. Recall that if $\psi \in V(I)$ and $g \in \mathbf{M}(I)$ then we say that (ψ,g) is a **reduction** of $f \in \mathbf{M}(I)$ if $\psi \circ f = g \circ \psi$. We have already seen one important example of a reduction in Section 6. Let (ψ,g) be a reduction of $f \in \mathbf{M}(I)$; then in a certain sense g describes the behaviour of f on $\mathrm{supp}(\psi)$, where

$$\mathrm{supp}(\psi) = \{ x \in I : \psi(J) \text{ is non-trivial for each open interval } J \subset I \text{ with } x \in J \} .$$

Thus the properties of g will give us information about f on $\mathrm{supp}(\psi)$. In this section we study the general properties of the reductions of a mapping $f \in \mathbf{M}(I)$. The results we obtain here will then be used in the following sections.

For $f \in \mathbf{M}(I)$ put $S(f) = T(f) \cup \{a,b\}$. Let $A \subset I$; we say that A is **f-invariant** if $f(A) \subset A$ and **f-almost-invariant** if $f(A-S(f)) \subset A$. A subset $D \subset I$ is called **perfect** if it is closed and contains no isolated points. Note that if $D \subset I$ is perfect then D is f-almost-invariant if and only if it is f-invariant. For $f \in \mathbf{M}(I)$ let $\mathbf{D}(f)$ denote the set of non-empty perfect subsets D of I such that D and $I-D$ are both f-almost-invariant. In Theorem 7.4 we show that if $f \in \mathbf{M}(I)$ and $D \subset I$ then there exists a reduction (ψ,g) of f with $\mathrm{supp}(\psi) = D$ if and only if $D \in \mathbf{D}(f)$. Theorem 7.4 also states that if (ψ,g) and (ψ',g') are two reductions of f with $\mathrm{supp}(\psi') \subset \mathrm{supp}(\psi)$ then there exists a unique $\theta \in V(I)$ such that $\psi' = \theta \circ \psi$, and $\theta \circ g = g' \circ \theta$; moreover, if $\mathrm{supp}(\psi') = \mathrm{supp}(\psi)$ then θ is a homeomorphism (and so in particular g and g' are conjugate). This last part of Theorem 7.4 implies that if (ψ,g) is a reduction of $f \in \mathbf{M}(I)$ then g is already determined up to conjugacy by $\mathrm{supp}(\psi)$.

In order to apply Theorem 7.4 we need perfect subsets of I . However, there is a standard method of obtaining these: If A is a closed subset of I then let

$$\kappa(A) = \{ x \in I : \text{each neighbourhood of } x \text{ contains uncountably many elements of } A \} ;$$

$\kappa(A)$ is called the set of **condensation points** of A . We have $\kappa(A) \subset A$ and by the Cantor-Bendixson theorem (see Lemma 7.5) $\kappa(A)$ is perfect and $A-\kappa(A)$ is countable. In particular, if A is uncountable then $\kappa(A)$ is also non-empty. ($\kappa(A)$ is really the only reasonable approximation to A by a perfect set.) Let $f \in \mathbf{M}(I)$; we say that $A \subset I$ is **f-weakly-invariant** if $f(A)-A$ is countable. Now let $f \in \mathbf{M}(I)$ and A be a closed uncountable subset of I ; in Lemma 7.6 we show that $\kappa(A) \in \mathbf{D}(f)$ if and only if A and $I-A$ are both f-weakly-invariant.

Putting together Theorem 7.4 and Lemma 7.6 we have Theorem 7.7: If $f \in \mathbf{M}(I)$ and A is a closed uncountable subset of I then there exists a reduction (ψ,g) of f with $\mathrm{supp}(\psi) = \kappa(A)$ if and only if A and $I-A$ are both f-weakly-invariant. The reductions which we consider in the present and later sections will all be obtained using this result. For example, let $f \in \mathbf{M}(I)$ with $I-Z(f)$ uncountable, then there exists a reduction (ψ,g) of f with $\mathrm{supp}(\psi) = \kappa(I-Z(f))$, because by Lemma 2.1 $Z(f)$ is f-invariant and it is easy to see that $I-Z(f)$ is f-almost-invariant. We therefore have $\mathrm{supp}(\psi) \subset I-Z(f)$ and $(I-Z(f)) - \mathrm{supp}(\psi)$ is countable, and Proposition 7.9 will show that $Z(g) = \varnothing$. Thus (ψ,g) essentially "kills off" the set $Z(f)$; in many cases g will provide information about the iterates of f which was "obscured" by $Z(f)$.

Another application of Theorem 7.7 is when we have $f \in \mathbf{M}(I)$ and an f-cycle C . If $I-A(C,f)$ is uncountable then there exists a reduction (ψ,g) of f with $\mathrm{supp}(\psi) = \kappa(I-A(C,f))$, because $I-A(C,f)$ is f-invariant and $A(C,f)$ is f-almost-invariant. Most of the reductions which occur in Section 8 will be obtained in this manner.

We now start our study of the reductions of a mapping $f \in \mathbf{M}(I)$ by looking at the properties of $\mathrm{supp}(\psi)$ when (ψ,g) is a reduction of f .

Proposition 7.1 (1) $\mathrm{supp}(\psi)$ is a non-empty perfect subset of I for each $\psi \in V(I)$.

(2) If D is a non-empty perfect subset of I then there exists $\psi \in V(I)$ with $\mathrm{supp}(\psi) = D$.

(3) Let ψ , $\psi' \in V(I)$ with $\mathrm{supp}(\psi') \subset \mathrm{supp}(\psi)$. Then there exists a unique $\theta \in V(I)$ such that $\psi' = \theta \circ \psi$. Moreover, if $\mathrm{supp}(\psi') = \mathrm{supp}(\psi)$ then θ is strictly increasing (and so a homeomorphism).

Proof (1): This is clear.

(2): This is a classical result in real analysis, and can be found, for example, in Carathéodory (1918). We give a proof in Section 13 (Proposition 13.5).

(3): Since $\mathrm{supp}(\psi') \subset \mathrm{supp}(\psi)$ it is easy to see that $\psi'(u) = \psi'(v)$ whenever $\psi(u) = \psi(v)$. We can thus define $\theta : I \to I$ by letting $\theta(x) = \psi'(y)$, where y is such that $\psi(y) = x$. Then $\psi' = \theta \circ \psi$, and θ is the only mapping with this property; also θ is increasing. Let $z \in I$; then we can find $y \in I$ with $\psi'(y) = z$, and so $\theta(\psi(y)) = z$. Thus θ is onto, and therefore continuous (since it is increasing). Hence $\theta \in V(I)$. The last part is clear. □

Let $f \in \mathbf{M}(I)$ and $\psi \in V(I)$; we say that ψ is **f-compatible** if for each open interval $J \subset I$ we have $\psi(J)$ is non-trivial if and only if $\psi(f(J))$ is non-trivial.

Proposition 7.2 Let $f \in \mathbf{M}(I)$ and $\psi \in V(I)$ be f-compatible. Then there exists a unique mapping $g : I \to I$ such that $\psi \circ f = g \circ \psi$, and in fact $g \in \mathbf{M}(I)$. Conversely, suppose (ψ, g) is a reduction of $f \in \mathbf{M}(I)$; then ψ is f-compatible.

Proof We first consider the converse, so let (ψ, g) be a reduction of $f \in \mathbf{M}(I)$ and $J \subset I$ be an open interval. Then $\psi(J)$ is non-trivial if and only if $(g \circ \psi)(J)$ is non-trivial (since $g \in \mathbf{M}(I)$). But $(g \circ \psi)(J) = (\psi \circ f)(J) = \psi(f(J))$, and thus ψ is f-compatible. Now suppose that $f \in \mathbf{M}(I)$ and $\psi \in V(I)$ is f-compatible. It follows that $\psi(f(x)) = \psi(f(y))$ whenever $\psi(x) = \psi(y)$, and so we can define $g : I \to I$ by letting $g(x) = \psi(f(y))$, where y is such that $x = \psi(y)$. Clearly we have $\psi \circ f = g \circ \psi$, and g is the only mapping with this property. Let $I_0, \ldots, I_N$

be the laps of f and let $M = \{ 0 \le k \le N : \psi(I_k)$ is non-trivial $\}$; then $I = \bigcup_{k \in M} \psi(I_k)$, and thus in order to show that $g \in \mathbf{M}(I)$ it is enough to show that g is continuous and strictly monotone on $\psi(I_k)$ for each $k \in M$. Let $k \in M$, and without loss of generality we can assume that f is increasing on I_k ; take $x , y \in \psi(I_k)$ with $x < y$ and let $u , v \in I_k$ be such that $x = \psi(u)$, $y = \psi(v)$ (so $u < v$). Then $\psi((u,v)) \supset (x,y)$ is non-trivial, and hence $\psi(f((u,v)))$ is also non-trivial. Since f is strictly increasing on $[u,v]$ this gives us that $\psi(f(v)) > \psi(f(u))$. But $\psi(f(v)) = g(\psi(v)) = g(y)$ and $\psi(f(u)) = g(\psi(u)) = g(x)$, and therefore g is strictly monotone on $\psi(I_k)$. Now $g(\psi(I_k)) = \psi(f(I_k))$, which is an interval; thus g must be continuous on $\psi(I_k)$ (since it is monotone there). □

Remark: Let (ψ,g) be a reduction of $f \in \mathbf{M}(I)$; Proposition 7.2 shows in particular that g is uniquely determined by f and ψ .

Let $f \in \mathbf{M}(I)$ and $A \subset I$; as already stated at the beginning of the section we say that A is f-invariant if $f(A) \subset A$ and f-almost-invariant if $f(A-S(f)) \subset A$, where $S(f) = T(f) \cup \{a,b\}$. Note that if $D \subset I$ is perfect then D is f-almost-invariant if and only if it is f-invariant.

Proposition 7.3 Let $f \in \mathbf{M}(I)$ and $\psi \in \mathbf{V}(I)$ with $\mathrm{supp}(\psi) = D$. Then ψ is f-compatible if and only if D and $I-D$ are both f-almost-invariant.

Proof It is easy to see that in the definition of ψ being f-compatible it is enough to consider open intervals $J \subset I$ with $J \cap S(f) = \emptyset$. Thus ψ is f-compatible if and only if for each open interval $J \subset I$ with $J \cap S(f) = \emptyset$ we have $J \cap D \neq \emptyset$ if and only if $f(J) \cap D \neq \emptyset$. On the other hand, let $A \subset I$ be closed; then A is f-almost-invariant if and only if $f(J) \cap A \neq \emptyset$ for each open interval $J \subset I$ with $J \cap S(f) = \emptyset$ and $J \cap A \neq \emptyset$, and $I-A$ is f-almost-invariant if and only if $J \cap A \neq \emptyset$ for each open interval $J \subset I$ with $J \cap S(f) = \emptyset$ and $f(J) \cap A \neq \emptyset$. Therefore ψ is f-compatible if and only if D and $I-D$ are f-almost-invariant. □

For $f \in \mathbf{M}(I)$ let $\mathbf{D}(f)$ denote the set of non-empty perfect subsets D of

I such that D and $I-D$ are both f-almost-invariant.

Theorem 7.4 Let $f \in \mathbf{M}(I)$ and $D \subset I$; then there exists a reduction (ψ,g) of f with $\mathrm{supp}(\psi) = D$ if and only if $D \in \mathbf{D}(f)$. If (ψ,g) and (ψ',g') are two reductions of f with $\mathrm{supp}(\psi') \subset \mathrm{supp}(\psi)$ then there exists a unique $\theta \in V(I)$ such that $\psi' = \theta \circ \psi$, and $\theta \circ g = g' \circ \theta$; moreover, if $\mathrm{supp}(\psi') = \mathrm{supp}(\psi)$ then θ is a homeomorphism (and so in particular g and g' are conjugate).

Proof This is just Propositions 7.1, 7.2 and 7.3 put together. (Note that $\theta \circ g = g' \circ \theta$ follows automatically from having $\psi' = \theta \circ \psi$, because θ is onto.) □

We next consider how to obtain elements of $\mathbf{D}(f)$. If A is a closed subset of I then $\kappa(A)$ denotes the set of condensation points of A ; i.e.

$$\kappa(A) = \{ x \in I : \text{each neighbourhood of } x \text{ contains uncountably many elements of } A \} .$$

We have $\kappa(A) \subset A$ and if $B \subset A$ is closed then $\kappa(B) \subset \kappa(A)$. Of course, if A is countable then $\kappa(A) = \emptyset$.

Lemma 7.5 Let A be a closed uncountable subset of I . Then:

(1) $\kappa(A)$ is a non-empty perfect subset of I .

(2) $A - \kappa(A)$ is countable.

(3) If $U \subset I$ is open then $U \cap \kappa(A) \neq \emptyset$ if and only if $U \cap A$ is uncountable.

(4) $\kappa(A) = A$ if and only if A is perfect.

Proof (2): Let $\{U_n\}_{n\geq 1}$ be a countable base for the topology on I , and let $N = \{ n \geq 1 : U_n \cap A \text{ is countable} \}$. Then we have $A - \kappa(A) = \bigcup_{n\in N} (U_n \cap A)$, which is countable.

(1): (2) implies that $\kappa(A)$ is non-empty, and by definition $\kappa(A)$ is closed. Let $x \in \kappa(A)$ and U be a neighbourhood of x ; then $U \cap A$ is uncountable, and so by (2) $U \cap \kappa(A)$ is uncountable. Thus $\kappa(A)$ contains no isolated points.

(3): This follows immediately from (2).

(4): If $\kappa(A) = A$ then by (1) A is perfect. Conversely, suppose that A is perfect. Then for each $x \in A$ we have that $A - \{x\}$ is (in the relative topology) a dense open subset of A. Thus if $U \subset I$ is open then by the Baire category theorem $U \cap A$ is either empty or uncountable. Therefore $\kappa(A) = A$. □

Remark: Parts (1) and (2) of Lemma 7.5 are called the Cantor-Bendixson Theorem. If A is a closed subset of I then $\kappa(A)$ is the only reasonable approximation to A by a perfect set. In fact, if $D \subset I$ is perfect and $D \neq \kappa(A)$ then $(D-A) \cup (A-D)$ is uncountable. This follows from Lemma 7.5 (2), since if D_1 and D_2 are perfect subsets of I with $D_1 \neq D_2$ then $(D_1-D_2) \cup (D_2-D_1)$ is uncountable.

Let $f \in M(I)$; recall that $A \subset I$ is said to be f-weakly-invariant if $f(A) - A$ is countable.

Lemma 7.6 Let $f \in M(I)$ and A be a closed uncountable subset of I. Then:

(1) $\kappa(A)$ is f-invariant (or, equivalently, $\kappa(A)$ is f-almost-invariant) if and only if A is f-weakly-invariant.

(2) $I - \kappa(A)$ is f-almost-invariant if and only if $I - A$ is f-weakly-invariant.

Proof (1): If $\kappa(A)$ is f-invariant then $f(\kappa(A)) \subset A$, and so $f(A) \subset A \cup f(A-\kappa(A))$. Hence by Lemma 7.5 (2) A is f-weakly-invariant. Conversely, suppose A is f-weakly-invariant; let $x \in \kappa(A)$ and U be a neighbourhood of $f(x)$. Then $f^{-1}(U)$ is a neighbourhood of x and so $f^{-1}(U) \cap A$ is uncountable. Therefore $U \cap A$ is also uncountable (since f is finite to one and $f(A)-A$ is countable), and thus $f(x) \in \kappa(A)$. Hence $\kappa(A)$ is f-invariant.

(2): If $I - \kappa(A)$ is f-almost-invariant then

$$\begin{aligned} f(I-A) \subset f(I-\kappa(A)) &\subset f((I-\kappa(A))-S(f)) \cup f(S(f)) \\ &\subset (I-\kappa(A)) \cup f(S(f)) \subset (I-A) \cup (A-\kappa(A)) \cup f(S(f)) . \end{aligned}$$

Hence by Lemma 7.5 (2) $I-A$ is f-weakly-invariant. Conversely, suppose that $I-A$ is f-weakly-invariant, and let $x \in (I-\kappa(A))-S(f)$. Then there exists a neighbourhood U of x such that $U \cap A$ is countable. But $f(U)$ is a neighbourhood of $f(x)$ (since f is monotone on $(x-\varepsilon,x+\varepsilon)$ for some $\varepsilon > 0$), and

$f(U) \cap A$ is countable, because

$$f(U) \cap A = (f(U-A) \cup f(U \cap A)) \cap A \subset (f(I-A)-(I-A)) \cup f(U \cap A) .$$

Therefore $f(x) \in I-\kappa(A)$, and hence $I-\kappa(A)$ is f-almost-invariant. □

Theorem 7.7 Let $f \in M(I)$ and A be a closed uncountable subset of I. Then there exists a reduction (ψ,g) of f with $\mathrm{supp}(\psi) = \kappa(A)$ if and only if A and $I-A$ are both f-weakly-invariant.

Proof This follows immediately from Theorem 7.4 and Lemmas 7.5 and 7.6. □

We can apply Theorem 7.7 with $A = I-Z(f)$: By Lemma 2.1 $Z(f)$ is f-invariant, and it is easily checked that $I-Z(f)$ is f-almost-invariant. Thus if $f \in M(I)$ and $I-Z(f)$ is uncountable then there exists a reduction (ψ,g) of f with $\mathrm{supp}(\psi) = \kappa(I-Z(f))$. (Hence $\mathrm{supp}(\psi) \subset I-Z(f)$ and $(I-Z(f)) - \mathrm{supp}(\psi)$ is countable.)

We next study the properties of g when (ψ,g) is a reduction of $f \in M(I)$.

Theorem 7.8 Let (ψ,g) be a reduction of $f \in M(I)$. Then:

(1) $T(g) \subset \psi(T(f))$, and so in particular $|T(g)| \le |T(f)|$.

(2) Let C be an f-cycle with period m and suppose that $\mathrm{int}(C) \cap \mathrm{supp}(\psi) \ne \emptyset$; then $\psi(C)$ is a g-cycle with period either m or $m/2$. Moreover, if C is proper then so is $\psi(C)$.

(3) If C is a topologically transitive f-cycle and $\mathrm{int}(C) \cap \mathrm{supp}(\psi) \ne \emptyset$ then $\psi(C)$ is topologically transitive.

(4) If R is an f-register-shift and $R \cap \mathrm{supp}(\psi) \ne \emptyset$ then $\psi(R)$ is a g-register-shift.

(5) If K is a g-cycle with period m then $\psi^{-1}(K)$ is an f-cycle with period m.

(6) $\overline{Z(g)} = \overline{\psi(Z(f) \cap \mathrm{supp}(\psi))}$; thus in particular $Z(g) = \emptyset$ if and only if $Z(f) \cap \mathrm{supp}(\psi) = \emptyset$.

Proof (1): Let $w \in T(g)$, and without loss of generality assume that w is a local maximum of g; let $\psi^{-1}(\{w\}) = [c,d]$ (and note that $[c,d] \subset (a,b)$). Then there exists $\varepsilon > 0$ such that f is increasing on $[c-\varepsilon,c]$ and is decreasing on $[d,d+\varepsilon]$; hence $[c,d] \cap T(f) \neq \emptyset$.

(2): Let $B_0,\ldots, B_{m-1}$ be the components of C. By assumption there exists j such that $\mathrm{supp}(\psi) \cap \mathrm{int}(B_j) \neq \emptyset$, and then $\psi(B_j)$ is a non-trivial closed interval. Also $g^n(\psi(B_k)) \subset \psi(B_i)$ whenever $f^n(B_k) \subset B_i$, and so in particular $\psi(B_k)$ is non-trivial for each k. Moreover, $\mathrm{int}(\psi(B_0)),\ldots, \mathrm{int}(\psi(B_{m-1}))$ are disjoint, and therefore by Lemma 3.9 $\psi(C)$ is a g-cycle with period either m or $m/2$. The last part is clear.

(3): Let $F \subset \psi(C)$ be closed with $g(F) \subset F$ and $\mathrm{int}(F) \neq \emptyset$; put $E = \psi^{-1}(F) \cap C$. Then E is a closed subset of C, $f(E) \subset E$ and $\mathrm{int}(E) \neq \emptyset$. Thus $E = C$ and so $\psi^{-1}(F) \supset C$. Hence $\psi(C) \subset \psi(\psi^{-1}(F)) = F$; i.e. $F = \psi(C)$.

(5): This is clear, since for any $A \subset I$ we have $f(\psi^{-1}(A)) \subset \psi^{-1}(g(A))$.

(6): Let $x \in Z(f) \cap \mathrm{supp}(\psi)$; then there exists $\varepsilon > 0$ such that $J = (x-\varepsilon,x+\varepsilon) \subset Z(f)$, and $\psi(J)$ is non-trivial because $x \in \mathrm{supp}(\psi)$. But g^n is clearly monotone on $\psi(J)$ for each $n \geq 0$, and hence $\psi(x) \in \overline{Z(g)}$. Thus $\overline{\psi(Z(f) \cap \mathrm{supp}(\psi))} \subset \overline{Z(g)}$. Conversely, let J_0 be a component of $Z(g)$ and for $n \geq 1$ let J_n be the component of $Z(g)$ with $g^n(J_0) \subset J_n$. Then either *(i)* the intervals $\{J_n\}_{n\geq 0}$ are disjoint, or *(ii)* J_k is a sink of g for some $k \geq 0$. Suppose first that *(i)* holds. Then the intervals $\{\psi^{-1}(J_n)\}_{n\geq 0}$ are also disjoint, and since $f^n(\psi^{-1}(J_0)) \subset \psi^{-1}(g^n(J_0)) \subset \psi^{-1}(J_n)$ for all $n \geq 0$, it follows that $\psi^{-1}(J_0)-Z(f)$ is finite. Now if $L \subset J_0$ is a non-trivial interval then $\psi^{-1}(L) \cap \mathrm{supp}(\psi)$ is uncountable, and hence $L \cap \psi(Z(f) \cap \mathrm{supp}(\psi)) \neq \emptyset$; i.e. $J_0 \subset \overline{\psi(Z(f) \cap \mathrm{supp}(\psi))}$. Suppose that *(ii)* holds: put $J = J_k$ (where $k \geq 0$ is such that J_k is a sink of g), and let $m \geq 1$ be such that $g^m(J) \subset J$ and g^m is increasing on J. Let $G = \{ x \in J : g^m(x) \neq x \}$; then G is open and $f^m(u) \neq u$ for all $u \in \psi^{-1}(G)$. If K is a component of G then $g^m(K) \subset K$ and so $f^m(\psi^{-1}(K)) \subset \psi^{-1}(K)$. We also have $f^m(\psi^{-1}(\{x\})) \subset \psi^{-1}(\{x\})$ for each $x \in J-G$, and thus if we let $N = \{ x \in J-G : \psi^{-1}(\{x\}) \cap T(f^m) \neq \emptyset \}$, then it is not hard to see that $\psi^{-1}(J-N)-Z_*(f)$ is finite. This gives us that $\psi^{-1}(J-N)-Z(f)$ is countable. Now let $L \subset J_0$ be a non-trivial interval; then $\psi(f^m(\psi^{-1}(L))) \supset g^m(L)$, and hence $f^m(\psi^{-1}(L)) \cap Z(f) \cap \mathrm{supp}(\psi)$ is uncountable. (Since N is finite we can find a non-trivial open interval $L' \subset g^m(L)-N$, and then $\psi^{-1}(L')-Z(f)$ is countable, $\psi^{-1}(L') \cap \mathrm{supp}(\psi)$ is uncountable and $\psi^{-1}(L') \subset f^m(\psi^{-1}(L))$.) But if $f^m(x) \in Z(f) \cap \mathrm{supp}(\psi)$ and $x \notin T(f^m)$ then $x \in Z(f) \cap \mathrm{supp}(\psi)$ (because $I - \mathrm{supp}(\psi)$ is f-almost-invariant). Therefore $\psi^{-1}(L) \cap Z(f) \cap \mathrm{supp}(\psi) \neq \emptyset$, and so $L \cap \psi(Z(f) \cap \mathrm{supp}(\psi)) \neq \emptyset$; i.e. $J_0 \subset \overline{\psi(Z(f) \cap \mathrm{supp}(\psi))}$.

(4): Let $\{K_n\}_{n\geq 1}$ be a generator for R ; then $\mathrm{int}(K_n) \cap \mathrm{supp}(\psi) \neq \emptyset$ for all $n \geq 1$. (We can find a component B of K_{n+2} with $B \subset \mathrm{int}(K_n)$, and $B \cap \mathrm{supp}(\psi) \neq \emptyset$ because $\mathrm{supp}(\psi)$ is f-invariant.) Thus by (2) $\psi(K_n)$ is a proper g-cycle with period either $\mathrm{per}(K_n)$ or $\mathrm{per}(K_n)/2$, and therefore $\{\psi(K_{2n})\}_{n\geq 1}$ is a splitting sequence of proper g-cycles. But $K_2 \cap Z(f) = \emptyset$, and so by (6) $\psi(K_2) \cap Z(g) = \emptyset$; hence $R' = \bigcap_{n\geq 1} \psi(K_{2n})$ is a g-register-shift. However, we have $\bigcap_{n\geq 1} \psi(K_{2n}) = \psi(\bigcap_{n\geq 1} K_{2n})$, since $\{K_{2n}\}_{n\geq 1}$ is a decreasing sequence of compact sets, and thus $R' = \psi(R)$. □

Let $\mathbf{M}_0(I) = \{ f \in \mathbf{M}(I) : Z(f) = \emptyset \}$. The following result is a corollary of Theorem 7.8.

Proposition 7.9 Let $f \in \mathbf{M}(I)$ with $I-Z(f)$ uncountable. Then there exists a reduction (ψ,g) of f with $\mathrm{supp}(\psi) = \kappa(I-Z(f))$, and we have:

(1) $g \in \mathbf{M}_0(I)$.

(2) If $C_1,\ldots, C_r$ are the topologically transitive f-cycles then $\psi(C_1),\ldots, \psi(C_r)$ are (r different) topologically transitive g-cycles. In particular, g has at least as many topologically transitive cycles as f .

(3) If $R_1,\ldots, R_\ell$ are the f-register-shifts then $\psi(R_1),\ldots, \psi(R_\ell)$ are (ℓ different) g-register-shifts. In particular, g has at least as many register-shifts as f .

(4) If (ψ',g') is a reduction of f with $g' \in \mathbf{M}_0(I)$ then there exists a unique $\theta \in \mathbf{V}(I)$ with $\psi' = \theta\circ\psi$; thus $\theta\circ g = g'\circ\theta$, and so (θ,g') is a reduction of g .

Proof We have already noted that there exists a reduction (ψ,g) of f with $\mathrm{supp}(\psi) = \kappa(I-Z(f))$.

(1): This follows from Theorem 7.8 (6).

(2): If C is a topologically transitive f-cycle then by Proposition 2.2 (3) $\mathrm{int}(C) \cap Z(f) = \emptyset$, and hence $\mathrm{int}(C) \cap \mathrm{supp}(\psi) \neq \emptyset$, since $\mathrm{int}(C)$ is uncountable. Thus by Theorem 7.8 (3) $\psi(C_1),\dots, \psi(C_r)$ are topologically transitive g-cycles. But $\mathrm{int}(C_1),\dots, \mathrm{int}(C_r)$ are disjoint and so $\mathrm{int}(\psi(C_1)),\dots, \mathrm{int}(\psi(C_r))$ are also disjoint. Therefore $\psi(C_1),\dots, \psi(C_r)$ are r different g-cycles.

(3): If R is an f-register-shift then R is an uncountable subset of $I-Z(f)$, and thus as in (2) we have that $\psi(R_1),\dots, \psi(R_\ell)$ are g-register-shifts. Now if A and B are disjoint subsets of I then it is easy to see that $\psi(A) \cap \psi(B)$ is countable. Hence by Proposition 2.2 (7) $\psi(R_1),\dots, \psi(R_\ell)$ are disjoint.

(4): If $Z(g') = \emptyset$ then by Theorem 7.8 (6) we have $Z(f) \cap \mathrm{supp}(\psi') = \emptyset$, and so $\mathrm{supp}(\psi') = \kappa(\mathrm{supp}(\psi')) \subset \kappa(I-Z(f)) = \mathrm{supp}(\psi)$. Thus Theorem 7.4 gives us the required $\theta \in V(I)$. □

Of course we can only apply Proposition 7.9 to $f \in M(I)$ if $I-Z(f)$ is uncountable. Note that $I-Z(f)$ will certainly be uncountable if there is either a topologically transitive f-cycle or an f-register-shift. If $I-Z(f)$ is countable then clearly $Z(f)$ is dense in I, but the converse is in general false, as the following simple example shows: Let $f \in M(I)$ have one turning point $\gamma \in (a,b)$, and suppose that $f(\gamma) = b$ and $f(b) = f(a) = a$. Hence f is increasing on $[a,\gamma]$ and decreasing on $[\gamma,b]$. Now I is the only f-cycle containing γ, and so in particular there can be no f-register-shift. Thus by Theorem 2.4 either $Z(f)$ is dense in I or I is a topologically transitive f-cycle (i.e. f is topologically transitive). It is easy to arrange that f is not topologically transitive (for example by giving f a fixed point in (a,γ)), and then $\overline{Z(f)}$ must be dense in I. (See the picture on the next page.) But $I-Z(f) = \{a,b\} \cup \bigcup_{m\geq 1} T(f^m)$, and $T(f^m)$ consists of 2^m-1 points such that between each two neighbouring points of $T(f^m)$ there is an element of $T(f^{m+1})$. Thus $I-Z(f)$ is uncountable. Let us apply Proposition 7.9 to this mapping f; we then get a reduction (ψ,g) of f with $g \in M_0(I)$. We have $g(b) = g(a) = a$ and $g(\psi(\gamma)) = \psi(f(\gamma)) = b$; hence $\gamma' = \psi(\gamma)$ is the (only) turning point of g, and I is the only g-cycle containing γ'. The same argument as before now gives us that g is topologically transitive (because $Z(g) = \emptyset$). This shows that the mapping g in Proposition 7.9 can have more topologically transitive cycles than f; in Section 10 we also give examples where g has more register-shifts than f .

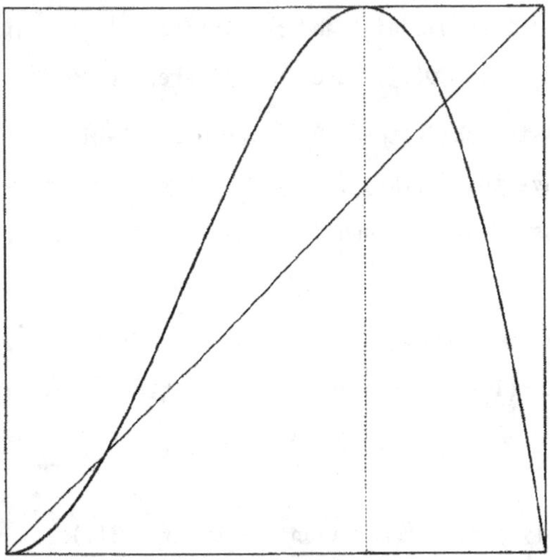

Let $f \in \mathbf{M}(I)$ and (ψ,g) be a reduction of f ; we next show that there is a natural bijection between $\mathbf{D}(g)$ and $\{ D \in \mathbf{D}(f) : D \subset \mathrm{supp}(\psi) \}$. This will be used in Section 8 in the analysis of the structure of $\mathbf{D}(f)$ for a mapping $f \in \mathbf{M}_0(I)$.

Lemma 7.10 Let (ψ,g) be a reduction of $f \in \mathbf{M}(I)$ and let (θ,g') be a reduction of g . Put $\psi' = \theta\circ\psi$, so (ψ',g') is also a reduction of f . Then $\mathrm{supp}(\psi') \subset \mathrm{supp}(\psi) \cap \psi^{-1}(\mathrm{supp}(\theta))$ and $[\mathrm{supp}(\psi) \cap \psi^{-1}(\mathrm{supp}(\theta))] - \mathrm{supp}(\psi')$ is countable; in particular $\mathrm{supp}(\psi') = \kappa(\mathrm{supp}(\psi) \cap \psi^{-1}(\mathrm{supp}(\theta)))$.

Proof Let $U = (I-\mathrm{supp}(\psi)) \cup \psi^{-1}(I-\mathrm{supp}(\theta))$ and $U' = I-\mathrm{supp}(\psi')$; thus $\mathrm{supp}(\psi') = I-U'$ and $\mathrm{supp}(\psi) \cap \psi^{-1}(\mathrm{supp}(\theta)) = I-U$. Now it is clear that $U \subset U'$; also if $x \in U'-U$ then it is easily checked that x is an end-point of $\overline{V}$ for some component V of $I-\mathrm{supp}(\psi)$, and hence $U'-U$ is countable. □

Let (ψ,g) be a reduction of $f \in \mathbf{M}(I)$, and let $D \in \mathbf{D}(g)$. By Theorem 7.4 there exists a reduction (θ,g') of g with $\mathrm{supp}(\theta) = D$; hence by Lemma 7.10 $\kappa(\mathrm{supp}(\psi) \cap \psi^{-1}(D)) = \mathrm{supp}(\theta\circ\psi)$, and so in particular $\kappa(\mathrm{supp}(\psi) \cap \psi^{-1}(D)) \in \mathbf{D}(f)$.

We can therefore define a mapping $\hat{\psi} : \mathbf{D}(g) \to \mathbf{D}(f)$ by $\hat{\psi}(D) = \kappa(\mathrm{supp}(\psi) \cap \psi^{-1}(D))$.

Proposition 7.11 Let (ψ,g) be a reduction of $f \in \mathbf{M}(I)$; then $\hat{\psi}$ maps $\mathbf{D}(g)$ bijectively onto $\{ D \in \mathbf{D}(f) : D \subset \mathrm{supp}(\psi) \}$.

Proof By definition we have $\hat{\psi}(D) \subset \mathrm{supp}(\psi)$ for each $D \in \mathbf{D}(g)$. Let D_1 , $D_2 \in \mathbf{D}(g)$ with $D_1 \neq D_2$, and for $i = 1$, 2 let $E_i = \mathrm{supp}(\psi) \cap \psi^{-1}(D_i)$. Then $(E_1-E_2) \cup (E_2-E_1)$ is uncountable, because $(D_1-D_2) \cup (D_2-D_1)$ is uncountable and $\mathrm{supp}(\psi) \cap \psi^{-1}(\{x\}) \neq \emptyset$ for each $x \in I$, and hence $\kappa(E_1) \neq \kappa(E_2)$, i.e. $\hat{\psi}(D_1) \neq \hat{\psi}(D_2)$; thus $\hat{\psi}$ is injective. Now let $D \in \mathbf{D}(f)$ with $D \subset \mathrm{supp}(\psi)$; by Theorem 7.4 there then exists a reduction (ψ',g') of f with $\mathrm{supp}(\psi') = D$ and $\theta \in \mathbf{V}(I)$ with $\psi' = \theta\circ\psi$. Therefore by Lemma 7.10 we have $\hat{\psi}(\mathrm{supp}(\theta)) = D$, and $\mathrm{supp}(\theta) \in \mathbf{D}(g)$ because (θ,g') is a reduction of g . □

Remark: Let (ψ,g) , (ψ',g') be reductions of $f \in \mathbf{M}(I)$ with $\mathrm{supp}(\psi) = \mathrm{supp}(\psi')$. By Theorem 7.4 there then exists a homeomorphism $\theta \in \mathbf{V}(I)$ with $\psi' = \theta\circ\psi$ and it is easily checked that $\hat{\psi}(D) = \hat{\psi}'(\theta(D))$ for each $D \in \mathbf{D}(g)$.

Note that $\hat{\psi}$ is order-preserving: If D , $D' \in \mathbf{D}(g)$ then $D \subset D'$ if and only if $\hat{\psi}(D) \subset \hat{\psi}(D')$.

In Section 8 we will need the following general fact about mappings in $\mathbf{M}_0(I)$:

Proposition 7.12 Let $f \in \mathbf{M}_0(I)$ and (ψ,g) be a reduction of f . Then we have:

(1) $g \in \mathbf{M}_0(I)$.

(2) If R' is a g-register-shift then $R' = \psi(R)$ for some f-register-shift R . In particular, f has at least as many register-shifts as g .

Proof (1): This follows immediately from Theorem 7.8 (6).

(2): Let R' be a g-register-shift and $\{K'_n\}_{n\geq 1}$ a generator for R' . For $n \geq 1$ let $K_n = \psi^{-1}(K'_n)$; then by Theorem 7.8 (5) $\{K_n\}_{n\geq 1}$ is a splitting sequence of f-cycles, and so by Proposition 2.11 $R = \bigcap_{n\geq 1} K_n$ is an f-register-shift, because $Z(f) = \emptyset$. But $\psi(K_n) = K'_n$ for each $n \geq 1$ and $\{K_n\}_{n\geq 1}$ is a decreasing sequence of compact sets; hence $\psi(R) = \bigcap_{n\geq 1} \psi(K_n) = R'$. □

There is no result corresponding to Proposition 7.12 (2) for topologically transitive cycles. Consider $f \in M_0(I)$ with a single f-register-shift and no topologically transitive f-cycles; if (ψ,g) is a reduction of f with $\text{supp}(\psi) \neq I$ then we will see in Lemma 8.15 that there are no g-register-shifts; thus by Proposition 7.12 (1) and Theorem 2.4 there is at least one topologically transitive g-cycle and so in particular g has more topologically transitive cycles than f . This can also occur without register-shifts being involved as the following mapping shows:

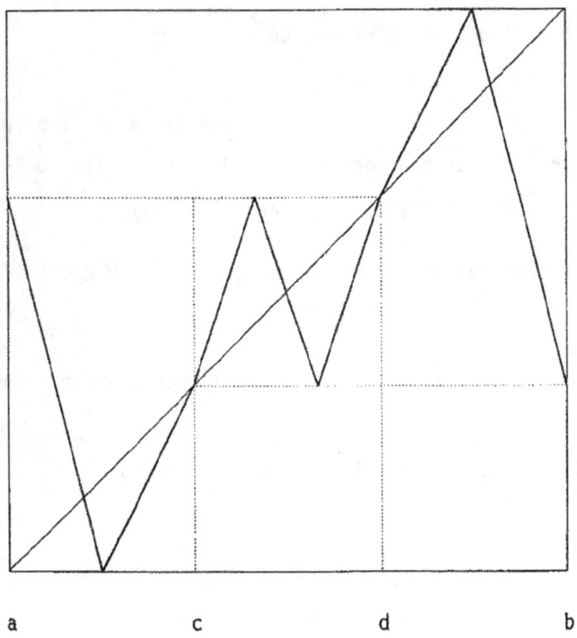

Here $[c,d]$ is a topologically transitive f-cycle, and it is not hard to see that $A([c,d],f)$ is dense in I ; thus $f \in M_0(I)$ and there is one topologically transitive f-cycle. Let $E = I - A([c,d],f)$; then E is closed and f-invariant and $I-E$ is f-almost-invariant; moreover, it is easily checked that E is uncountable, and hence by Theorem 7.7 there exists a reduction (ψ,g) of f with $\text{supp}(\psi) = \kappa(E)$. This reduction "kills off" the interval (c,d) and all that gets mapped into it; g looks something like:

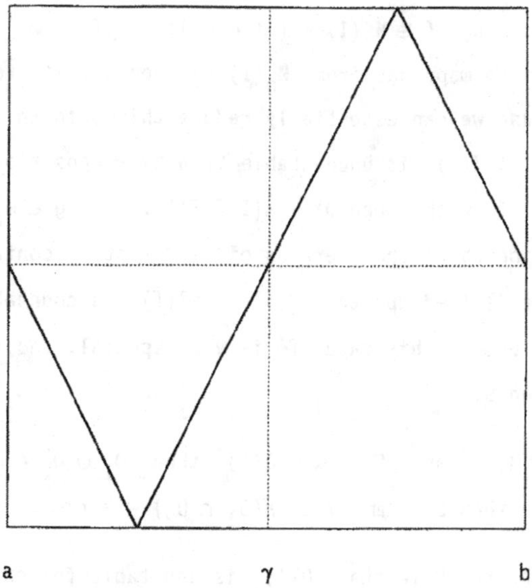

In particular there are two topologically transitive g-cycles ($[a,\gamma]$ and $[\gamma,b]$), and so g has more topologically transitive cycles than f .

Remark: In the above example we considered a reduction (ψ,g) of f with $\mathrm{supp}(\psi) = \kappa(I - A([c,d],f))$. More generally, if $f \in \mathbf{M}(I)$ and C is any f-cycle then $I - A(C,f)$ is f-invariant and $A(C,f)$ is f-almost-invariant. Thus if $I - A(C,f)$ is uncountable then by Theorem 7.7 there exists a reduction (ψ,g) of f with $\mathrm{supp}(\psi) = \kappa(I - A(C,f))$. Most of the reductions which occur in Section 8 will be obtained in this manner.

8. THE STRUCTURE OF THE SET D(f)

In this section we continue the study of reductions by examining the structure of the set $\mathbf{D}(f)$ for a mapping $f \in \mathbf{M}_0(I) = \{ f \in \mathbf{M}(I) : Z(f) = \varnothing \}$. The reason for making this restriction to mappings from $\mathbf{M}_0(I)$ is because it greatly simplifies the analysis, and because we can essentially reduce things to this case: If $f \in \mathbf{M}(I)$ is such that $I-Z(f)$ is uncountable then by Proposition 7.9 there exists a reduction (ψ,g) of f with $\mathrm{supp}(\psi) = \kappa(I-Z(f))$, and $g \in \mathbf{M}_0(I)$. Now most of the important information about the iterates of f is still contained in g, and this information can be "lifted up" to f. If $I-Z(f)$ is countable then we cannot apply Proposition 7.9; but in this case f is very special, and such mappings will be dealt with in Section 9.

Note that if $f \in \mathbf{M}(I)$ and $D_1, D_2 \in \mathbf{D}(f)$ then $D_1 \cup D_2 \in \mathbf{D}(f)$; also if $D_1 \cap D_2$ is uncountable then by Lemma 7.6 $\kappa(D_1 \cap D_2) \in \mathbf{D}(f)$.

We will show in Theorem 8.10 that $\mathbf{D}(f)$ is countable for each $f \in \mathbf{M}_0(I)$, and that $\mathbf{D}(f)$ is finite if and only if each f-register-shift is tame. In particular, $\mathbf{D}(f)$ is finite for each $f \in \mathbf{M}_*(I)$, where $\mathbf{M}_*(I)$ denotes the set of mappings $f \in \mathbf{M}_0(I)$ for which there are no f-register-shifts. Note that if (ψ,g) is a reduction of $f \in \mathbf{M}_*(I)$ then by Proposition 7.12 we also have $g \in \mathbf{M}_*(I)$.

Let $f \in \mathbf{M}_*(I)$; it turns out that the maximal elements of the set $\{ D \in \mathbf{D}(f) : D \neq I \}$ are exactly the sets of the form $\kappa(I-A(C,f))$ with C a topologically transitive f-cycle. We call a reduction (ψ,g) of f primary if $\mathrm{supp}(\psi) = \kappa(I-A(C,f))$ for some topologically transitive f-cycle C. In Theorem 8.8 we show that if (ψ,g) is a reduction of $f \in \mathbf{M}_*(I)$ with $\mathrm{supp}(\psi) \neq I$ then there exists $n \geq 1$ and $(\psi_1,f_1),\ldots,(\psi_n,f_n)$ such that (ψ_k,f_k) is a primary reduction of f_{k-1} for $k = 1,\ldots,n$ (with $f_0 = f$) and $(\psi,g) = (\psi_n \circ \cdots \circ \psi_1, f_n)$. If $D \in \mathbf{D}(f)$ then by Theorem 7.4 $D = \mathrm{supp}(\psi)$ for some reduction (ψ,g) of f; thus Theorem 8.8 gives a procedure for constructing the elements of $\mathbf{D}(f)$ for a mapping $f \in \mathbf{M}_*(I)$. It follows from Theorem 8.8 that if $f \in \mathbf{M}_0(I)$ with $|T(f)| = 1$ then $\mathbf{D}(f)$ is linearly ordered by inclusion.

In Theorem 8.13 we extend Theorem 8.8 to mappings in $\mathbf{M}_0(I)$. For mappings in $\mathbf{M}_0(I)$ it is necessary to use two types of reductions to obtain the analogue of

Theorem 8.8: as well as primary reductions as defined above we also need what we will call register-shift reductions.

We start our analysis of $D(f)$ by considering the case of a mapping f from $M_*(I)$. First we need a couple of general facts about f-almost-invariant subsets of I .

Lemma 8.1 Let $f \in M(I)$ and $A \subset I$ be f-almost-invariant. Then A is also f^n-almost-invariant for each $n \geq 1$.

Proof Since $T(f^n) = \{ x \in I : f^k(x) \in T(f) \text{ for some } 0 \leq k < n \}$ and $f^{-1}(\{a,b\}) \subset S(f)$ we have $S(f^n) = \{ x \in I : f^k(x) \in S(f) \text{ for some } 0 \leq k < n \}$, and thus $S(f^n) = S(f) \cup f^{-1}(S(f^{n-1}))$. Therefore

$$f(A-S(f^n)) = f((A-S(f)) - f^{-1}(S(f^{n-1}))) \subset f(A-S(f)) - S(f^{n-1}) \subset A-S(f^{n-1})$$

(because $f(E-f^{-1}(F)) \subset f(E)-F$ for all $E , F \subset I$). Hence $f^n(A-S(f^n)) \subset f^{n-1}(A-S(f^{n-1}))$, and so by induction it follows that A is f^n-almost-invariant. □

Lemma 8.2 Let $f \in M(I)$ and $A \subset I$ be f-almost-invariant. Then $\bar{A}$ and $\text{int}(A)$ are both also f-almost-invariant.

Proof Let $x \in \bar{A}-S(f)$; then there exists a sequence $\{x_n\}_{n\geq1}$ from $A-S(f)$ with $x = \lim_{n\to\infty} x_n$. Since A is f-almost-invariant we have $f(x_n) \in A$ for all n , and thus $f(x) \in \bar{A}$, i.e. $f(\bar{A}-S(f)) \subset \bar{A}$. Hence $\bar{A}$ is f-almost-invariant. Now let $J \subset I-S(f)$ be an open interval; then f is monotone on J and $J \subset (a,b)$; thus $f(J)$ is an open interval. This shows that if $U \subset I$ is open then $f(U-S(f))$ is also open; in particular $f(\text{int}(A)-S(f))$ is open. Therefore $f(\text{int}(A)-S(f)) \subset \text{int}(A)$, since $f(\text{int}(A)-S(f)) \subset f(A-S(f)) \subset A$, and hence $\text{int}(A)$ is f-almost-invariant. □

Let $f \in M(I)$ and $U \subset I$ be open and f-almost-invariant. If J is a connected component of U then $f(J-S(f)) \subset U$; but $f(J-S(f))$ is open and connected, and so there exists a (unique) component K of U with $f(\bar{J}) \subset \bar{K}$. Now let $n \geq 1$; by

Lemma 8.1 U is f^n-almost-invariant and so the same argument shows that there exists a unique component L of U with $f^n(\bar{J}) \subset \bar{L}$. We say that a component J of U is **periodic** if $f^m(\bar{J}) \subset \bar{J}$ for some $m \geq 1$; the smallest such $m \geq 1$ is called the **period** of J. A component K of U is called **eventually periodic** if $f^n(\bar{K}) \subset \bar{J}$ for some periodic component J of U and some $n \geq 0$; we also say that m is the **eventual period** of K, where m is the period of J. (It is easy to see that this is well-defined.) Note that if J is a component of U which is not eventually periodic then the intervals $\{\mathrm{int}(f^n(J))\}_{n\geq 0}$ are disjoint, and from this it follows that $J-Z(f)$ is finite. Thus if $f \in \mathbf{M}_0(I)$ then each component of U is eventually periodic. Moreover, if $f \in \mathbf{M}_0(I)$ then U has only finitely many periodic components (since if J is a component of U then $f^n(J) \cap S(f) \neq \varnothing$ for some $n \geq 0$).

Lemma 8.3 Let $f \in \mathbf{M}(I)$, $D \in \mathbf{D}(f)$ and C be a topologically transitive f-cycle. Then either $A(C,f) \subset D$ or $A(C,f) \cap D = \varnothing$.

Proof Suppose $A(C,f) \cap (I-D) \neq \varnothing$; then $U = \mathrm{int}(C) \cap (I-D)$ is a non-empty f-almost-invariant open set. Consider a connected component J of U ; J must be periodic, since otherwise $F = \overline{\bigcup_{n\geq 1} f^n(J)}$ would be an f-invariant closed subset of C with $\mathrm{int}(F) \neq \varnothing$ and $F \cap J = \varnothing$. Let m be the period of J ; then $K = \bigcup_{j=0}^{m-1} f^j(\bar{J})$ is a closed f-invariant subset of C with $\mathrm{int}(K) \neq \varnothing$, and thus $K = C$. But $\mathrm{int}(C) \cap D \subset K-U$ and $K-U$ is finite. Therefore $\mathrm{int}(C) \cap D = \varnothing$, because D is perfect. Hence $A(C,f) \cap D = \varnothing$, since D is f-invariant. □

Theorem 8.4 If $f \in \mathbf{M}_*(I)$ then $\mathbf{D}(f)$ is finite.

Proof Let $f \in \mathbf{M}_*(f)$ with topologically transitive f-cycles $C_1,\dots, C_r$. For $M \subset \{1,\dots,r\}$ put $E_M = I - \bigcup_{j\in M} A(C_j,f)$; then E_M is closed and f-invariant and $I-E_M$ is f-almost-invariant. Thus, letting $D_M = \kappa(E_M)$, we have either $D_M = \varnothing$ or $D_M \in \mathbf{D}(f)$. Now let $D \in \mathbf{D}(f)$ and put $M = \{ 1 \leq j \leq r : A(C_j,f) \cap D = \varnothing \}$; we have $D \subset E_M$, and hence also $D \subset D_M$. On the other hand, if $j \notin M$ then by Lemma 8.3 $A(C_j,f) \subset D$, and so $D_M - D \subset E_M - D \subset I - \bigcup_{j=1}^{r} A(C_j,f)$. Thus by Theorem

2.4 $\mathrm{int}(D_M - D) = \varnothing$. The theorem therefore follows from the next result. □

Proposition 8.5 Let $f \in \mathbf{M}_0(I)$ and $D \in \mathbf{D}(f)$; then

$$\{ D' \in \mathbf{D}(f) : D' \subset D \text{ and } \mathrm{int}(D-D') = \varnothing \}$$

is finite.

Proof Fix $D \in \mathbf{D}(f)$ and put $W = \{ D' \in \mathbf{D}(f) : D' \subset D$ and $\mathrm{int}(D-D') = \varnothing \}$; if $D = I$ then $W = \{I\}$ and so we can assume that $D \neq I$. Let $U_1, \ldots, U_m$ be the periodic components of $I-D$; since each component of $I-D$ is eventually periodic we have $m \geq 1$. Let $D' \in W$; then $I-D \subset I-D'$ and so each periodic component of $I-D$ is contained in a periodic component of $I-D'$; we thus obtain a partition $\pi(D')$ of $\{1,2,\ldots,m\}$: j and k are in the same element of $\pi(D')$ if and only if U_j and U_k are subsets of the same periodic component of $I-D'$. Put

$$T(D') = \{ w \in T(f) : w \text{ is contained in some periodic component of } I-D' \} .$$

Now since there are only finitely many partitions of $\{1,2,\ldots,m\}$ and $T(f)$ is finite, it is enough to show that $D' \in W$ is uniquely determined by $\pi(D')$ and $T(D')$. Thus let $D', D'' \in W$ with $\pi(D') = \pi(D'')$ and $T(D') = T(D'')$; we will show that $D' = D''$. Let $P(D')$ (resp. $P(D'')$) denote the set of periodic components of $I-D'$ (resp. of $I-D''$). If $U' \in P(D')$ then $\mathrm{int}(U' - (I-D)) = \varnothing$; thus $U_k \subset U'$ for some k , because each component of $I-D$ is eventually periodic. Therefore if we define $\gamma' : P(D') \to \pi(D')$ by $\gamma'(U') = \{ 1 \leq k \leq m : U_k \subset U' \}$ then γ' is a bijection. In the same way we have a bijection $\gamma'' : P(D'') \to \pi(D'')$, and hence a bijection $\alpha = (\gamma'')^{-1} \circ \gamma' : P(D') \to P(D'')$. We can thus write $P(D') = \{U'_1,\ldots,U'_p\}$ and $P(D'') = \{U''_1,\ldots,U''_p\}$ so that $\alpha(U'_j) = U''_j$ for $j = 1,\ldots, p$. It follows that $U'_j \cap U''_j \neq \varnothing$ for each j (since $U_k \subset U'_j \cap U''_j$ for each $k \in \gamma''(U''_j)$), and also $U'_j \cap U''_i = \varnothing$ for all $j \neq i$. (If $U'_j \cap U''_i \neq \varnothing$ then, because $\mathrm{int}((U'_j \cap U''_i) - (I-D)) = \varnothing$ and each component of $I-D$ is eventually periodic, we have that $U_k \subset U'_j \cap U''_i$ for some k , and hence $k \in \gamma'(U'_j) \cap \gamma''(U''_i)$; this implies that $j = i$.) Suppose $U'_j \neq U''_j$ for some j ; then without loss of generality we can assume there exist $v < w$ with $(v,w] \subset U'_j$ and with w the left-hand end-point of $\overline{U''}$. (Note $U'_j - U''_j \subset \{a,b\}$ is not possible, since D' and D'' are perfect.) Now $T(D') = T(D'')$ and $U'_j \cap U''_i = \emptyset$ for all $i \neq j$, and so f is monotone on $[v,w+\varepsilon]$ for some $\varepsilon > 0$. Let k be such that $f(\overline{U'_j}) \subset \overline{U'_k}$; then it is easy to see that also $f(\overline{U''_j}) \subset \overline{U''_k}$. But $f(w)$ is an end-point of $\overline{U''_k}$ (because $f(w) \in D''$), and therefore $f((v,w]) \subset U'_k - U''_k$. Repeating this argument shows that for each $n \geq 0$ there exists q so that $f^n((v,w]) \subset U'_q - U''_q$. In particular f^n is monotone on $[v,w]$ for all $n \geq 0$, which contradicts the fact that $Z(f) = \emptyset$. Hence we must have $U'_j = U''_j$ for all j. From this it easily follows that $D' = D''$: Let $E' = E'' = \{ x \in I : f^n(x) \notin \bigcup_{j=1}^{p} U'_j$ for all $n \geq 0 \}$; then $D' \subset E'$ and $E'-D'$ is countable (because each component of $I-D'$ is eventually periodic), and in the same way $D'' \subset E''$ and $E''-D''$ is countable. Thus $(D'-D'') \cup (D''-D')$ is countable, and so $D' = D''$, since D' and D'' are both perfect. □

We now give a procedure for constructing the elements of $\mathbf{D}(f)$ for a mapping $f \in \mathbf{M}_*(I)$.

Note that if (ψ,g) is a reduction of $f \in \mathbf{M}_*(I)$ then by Proposition 7.12 we have $g \in \mathbf{M}_*(I)$. Let $f \in \mathbf{M}_*(I)$ and (ψ,g) be a reduction of f ; we call (ψ,g) **primary** if there exists a topologically transitive f-cycle C such that $\text{supp}(\psi) = \kappa(I-A(C,f))$. (If C' is any f-cycle then $I-A(C',f)$ is f-invariant and $A(C',f)$ is f-almost-invariant; thus if $I-A(C',f)$ is uncountable then by Theorem 7.7 there exists a reduction (ψ',g') of f with $\text{supp}(\psi') = \kappa(I-A(C',f))$. This implies that each topologically transitive f-cycle C with $I-A(C,f)$ uncountable gives rise to a primary reduction of f.)

Lemma 8.6 Let $f \in \mathbf{M}_*(I)$ and (ψ,g) be a primary reduction of f ; suppose there exists a reduction (ψ',g') of f and $\theta \in V(I)$ such that $\psi = \theta \circ \psi'$, and so the diagram on the next page commutes. Then either $\text{supp}(\psi') = I$, in which case ψ' is a homeomorphism and f and g' are conjugate, or $\text{supp}(\psi') = \text{supp}(\psi)$, in which case θ is a homeomorphism and g' and g are conjugate.

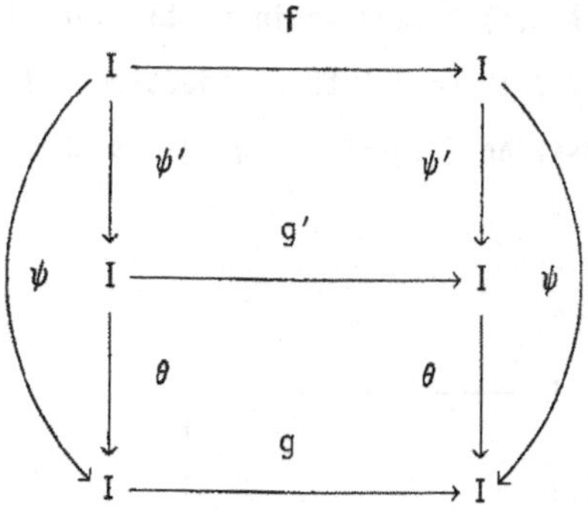

Proof Since $\psi = \theta \circ \psi'$ we have $\mathrm{supp}(\psi) \subset \mathrm{supp}(\psi')$. Let C be the topologically transitive f-cycle with $\mathrm{supp}(\psi) = \kappa(I-A(C,f))$; then by Lemma 8.3 either $A(C,f) \subset \mathrm{supp}(\psi')$ or $A(C,f) \cap \mathrm{supp}(\psi') = \emptyset$. If the former holds then $\mathrm{supp}(\psi') \supset A(C,f) \cup \kappa(I-A(C,f))$, and this means that $I-\mathrm{supp}(\psi')$ is countable; hence $\mathrm{supp}(\psi') = I$, and in this case ψ' is a homeomorphism. If the latter holds then $\mathrm{supp}(\psi') = \kappa(\mathrm{supp}(\psi')) \subset \kappa(I-A(C,f)) = \mathrm{supp}(\psi)$; thus $\mathrm{supp}(\psi') = \mathrm{supp}(\psi)$, and therefore by Proposition 7.1 (3) θ is a homeomorphism. □

Let (ψ,g) be a primary reduction of $f \in \mathbf{M}_*(I)$, and let $D \in \mathbf{D}(f)$ with $\mathrm{supp}(\psi) \subset D$. Then by Lemma 8.6 and Proposition 7.1 (3) we have either $D = I$ or $D = \mathrm{supp}(\psi)$; i.e. $\mathrm{supp}(\psi)$ is a maximal element of the set $\{ D \in \mathbf{D}(f) : D \neq I \}$. The next lemma shows in particular that the converse of this is true, namely that each maximal element of the set $\{ D \in \mathbf{D}(f) : D \neq I \}$ has the form $\mathrm{supp}(\psi)$ for some primary reduction (ψ,g) of f .

Lemma 8.7 Let $f \in \mathbf{M}_*(I)$ and (ψ',g') be a reduction of f with $\mathrm{supp}(\psi') \neq I$. Then there exists a primary reduction (ψ,g) of f and $\theta \in \mathbf{V}(I)$ such that $\psi' = \theta \circ \psi$, and hence the diagram on the following page commutes.

Proof Let $C_1,\ldots, C_r$ be the topologically transitive f-cycles; since $f \in \mathbf{M}_*(I)$ Theorem 2.4 implies that $A(C_1,f) \cup \cdots \cup A(C_r,f)$ is a dense subset of I . Thus we cannot have $A(C_j,f) \subset \mathrm{supp}(\psi')$ for all $j = 1,\ldots, r$ (because this would imply that $\mathrm{supp}(\psi') = I$). Therefore by Lemma 8.3 there exists a topologically transitive f-cycle C with $A(C,f) \cap \mathrm{supp}(\psi') = \emptyset$. Hence

$\mathrm{supp}(\psi') = \kappa(\mathrm{supp}(\psi')) \subset \kappa(I\text{-}A(C,f))$, and so in particular $I\text{-}A(C,f)$ is uncountable; thus by Theorem 7.7 there exists a reduction (ψ,g) of f with $\mathrm{supp}(\psi) = \kappa(I\text{-}A(C,f))$. Proposition 7.1 (3) now gives us that there exists $\theta \in V(I)$ such that $\psi' = \theta\circ\psi$. □

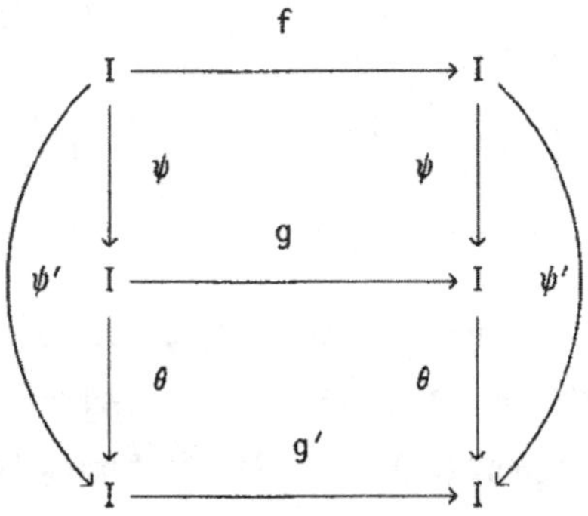

Let $f \in \mathbf{M}_*(I)$ and D be a maximal element of the set $\{ D' \in \mathbf{D}(f) : D' \neq I \}$. By Theorem 7.4 there exists a reduction (ψ',g') of f with $\mathrm{supp}(\psi') = D$, and since $\mathrm{supp}(\psi') \neq I$ Lemma 8.7 gives us a primary reduction (ψ,g) of f and $\theta \in \mathbf{V}(I)$ with $\psi' = \theta\circ\psi$. But then $\mathrm{supp}(\psi') \subset \mathrm{supp}(\psi)$, and so by the maximality of D we must have $D = \mathrm{supp}(\psi)$.

Theorem 8.8 Let $f \in \mathbf{M}_*(I)$ and (ψ,g) be a reduction of f with $\mathrm{supp}(\psi) \neq I$. Then there exists $n \geq 1$ and $(\psi_1,f_1),\ldots,(\psi_n,f_n)$ such that (ψ_k,f_k) is a primary reduction of f_{k-1} for $k = 1,\ldots, n$ (with $f_0 = f$) and $(\psi,g) = (\psi_n\circ\cdots\circ\psi_1,f_n)$.

Proof Suppose that for some $m \geq 1$ we have $(\psi_1,f_1),\ldots, (\psi_m,f_m)$ and $\theta_m \in V(I)$ such that (ψ_k,f_k) is a primary reduction of f_{k-1} for $k = 1,\ldots, m$ (with $f_0 = f$) and $\psi = \theta_m\circ\psi_m\circ\cdots\circ\psi_1$. We say then that $\{(\psi_1,f_1),\ldots,(\psi_m,f_m),\theta_m\}$ is an m-chain. Since $\mathrm{supp}(\psi) \neq I$ we have by Lemma 8.7 that a 1-chain exists. Let $\{(\psi_1,f_1),\ldots,(\psi_m,f_m),\theta_m\}$ be an m-chain, and for $k = 1,\ldots, m$ let $D_k = \mathrm{supp}(\psi_k\circ\cdots\circ\psi_1)$; then $D_k \in \mathbf{D}(f)$ and $D_m \subset D_{m-1} \subset \cdots \subset D_1$. Moreover, $D_k \neq D_{k+1}$ for each $k = 1,\ldots, m-1$ (since if $D_k = D_{k+1}$ then by Proposition 7.1 (3) ψ_{k+1} would be a homeomorphism), and therefore $m \leq |\mathbf{D}(f)|$. Suppose now that

θ_m is not a homeomorphism. Then (θ_m, g) is a reduction of f_m with $\mathrm{supp}(\theta_m) \neq I$, and by Proposition 7.12 $f_m \in \mathbf{M}_*(I)$; thus by Lemma 8.7 there exists a primary reduction (ψ_{m+1}, f_{m+1}) of f_m and $\theta_{m+1} \in V(I)$ such that $\theta_m = \theta_{m+1} \circ \psi_{m+1}$. Hence $\{(\psi_1, f_1), \ldots, (\psi_m, f_m), (\psi_{m+1}, f_{m+1}), \theta_{m+1}\}$ is an (m+1)-chain. But by Theorem 8.4 $\mathbf{D}(f)$ is finite, and so there must exist $n \geq 1$ and an n-chain $\{(\psi_1', f_1'), \ldots, (\psi_n', f_n'), \theta_n'\}$ such that θ_n' is a homeomorphism. Let $\psi_n = \theta_n' \circ \psi_n'$; then $\mathrm{supp}(\psi_n) = \mathrm{supp}(\psi_n')$, and so (ψ_n, g) is a primary reduction of f_{n-1}' . Now put $(\psi_k, f_k) = (\psi_k', f_k')$ for $k = 1, \ldots, n-1$ and $f_n = g$. Then (ψ_k, f_k) is a primary reduction of f_{k-1} for $k = 1, \ldots, n$ and $(\psi, g) = (\psi_n \circ \cdots \circ \psi_1, f_n)$. □

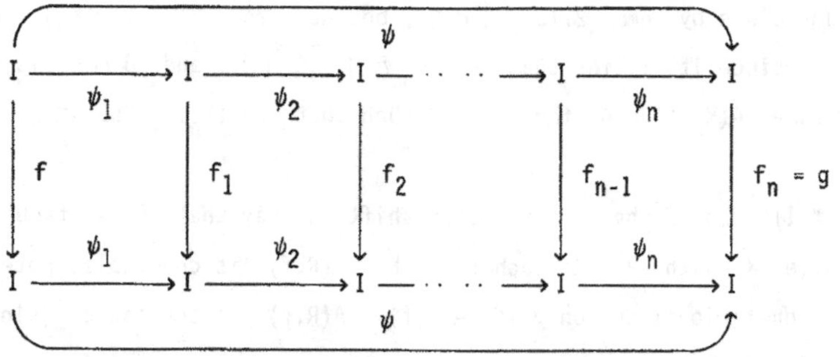

Let $f \in \mathbf{M}_*(I)$ and $D \in \mathbf{D}(f)$ with $D \neq I$. Then by Theorems 7.7 and 8.8 there exists $n \geq 1$ and $(\psi_1, f_1), \ldots, (\psi_n, f_n)$ such that (ψ_k, f_k) is a primary reduction of f_{k-1} for $k = 1, \ldots, n$ (with $f_0 = f$), and $D = \mathrm{supp}(\psi_n \circ \cdots \circ \psi_1)$.

We now extend the analysis we have given for mappings in $\mathbf{M}_*(I)$ to the mappings in $\mathbf{M}_0(I)$. First we need a result for register-shifts which corresponds to Lemma 8.3.

Lemma 8.9 Let $f \in \mathbf{M}(I)$, $D \in \mathbf{D}(f)$ and R be an f-register-shift; let $(K_n)_{n \geq 1}$ be a generator for R . Then either $R \subset D$, in which case $A(K_n, f) \subset D$ for all large enough n , or $R \cap D = \emptyset$, in which case $A(K_n, f) \cap D = \emptyset$ for all large enough n . In particular, either $A(R, f) \subset D$ or $A(R, f) \cap D = \emptyset$.

Proof If $R \cap D \neq \emptyset$ then $R \subset D$ follows immediately from Proposition 2.10 because D is closed and f invariant. If $R \cap D = \emptyset$ then there exists $\varepsilon > 0$ such that

$|z-x| > \varepsilon$ for all $z \in R$, $x \in D$, and hence $K_n \cap D = \emptyset$ for all large enough n. But if $K_n \cap D = \emptyset$ then $\mathrm{int}(K_n) \cap D = \emptyset$, and so $A(K_n,f) \cap D = \emptyset$, since D is f-invariant. Assume now that $R \subset D$, and let

$$p = \max\{ \text{ period of } J : J \text{ is a periodic component of } I-D \text{ with } J \cap S(f) \neq \emptyset \} .$$

Choose n so that $\mathrm{per}(K_n) > 2p$, and suppose that $A(K_n,f) \cap (I-D) \neq \emptyset$; then $\mathrm{int}(K_n) \cap (I-D) \neq \emptyset$, because $I-D$ is f-almost-invariant. Thus, since $K_n \cap Z(f) = \emptyset$, there exists a periodic component J of $I-D$ with $J \cap S(f) \neq \emptyset$ such that $\mathrm{int}(K_n) \cap J \neq \emptyset$. Let q be the period of J; by Lemmas 3.2 and 3.9 $Q = \bigcup_{k=0}^{q-1} f^k(\bar{J})$ is an f-cycle with period either q or $q/2$, and we have $K_n \cap Q \neq \emptyset$. Therefore by Lemma 2.15 $K_n \subset Q$, because $2q \le 2p < \mathrm{per}(K_n)$. But this is not possible, since it implies that $R \subset K_n \cap D \subset Q \cap D$, and $Q \cap D$ is finite. Hence we must have $A(K_n,f) \subset D$ for all n such that $\mathrm{per}(K_n) > 2p$. □

Let $f \in \mathbf{M}(I)$ and R be an f-register-shift; we say that R is **tame** if there exists an f-cycle K with $R \subset K$ such that $K - A(R,f)$ is countable. Note that $K - A(R,f)$ is countable if and only if $A(K,f) - A(R,f)$ is countable, (since $K - A(K,f)$ is finite and

$$A(K,f) - A(R,f) \subset \{ x \in I : f^n(x) \in K - A(R,f) \text{ for some } n \ge 0 \}).$$

If R is tame and $\{K_n\}_{n\ge1}$ is a generator for R then by Lemma 2.15 $A(K_n,f) - A(R,f)$ is countable for all large enough n.

We have seen in Section 5 that tame register-shifts can exist; however, in Section 9 we show that their structure is very special.

Theorem 8.10 $\mathbf{D}(f)$ is countable for each $f \in \mathbf{M}_0(I)$. Moreover, $\mathbf{D}(f)$ is finite if and only if each f-register-shift is tame.

Proof Let $f \in \mathbf{M}_0(I)$ with topologically transitive f-cycles $C_1,\dots,C_r$ and f-register-shifts $R_1,\dots,R_\ell$; for $1 \le i \le \ell$ let $\{K_n^i\}_{n\ge1}$ be a generator for R_i. For $M \subset \{1,\dots,r\}$, $N \subset \{1,\dots,\ell\}$ and $n \ge 1$ put $E_M = I - \bigcup_{j\in M} A(C_j,f)$, $G_N^n = I - \bigcup_{i\in N} A(K_n^i,f)$, and $D_{N,M}^n = \kappa(G_N^n \cap E_M)$. Thus either $D_{N,M}^n = \emptyset$ or

$D^n_{N,M} \in \mathbf{D}(f)$. Now let $D \in \mathbf{D}(f)$ and put $M = \{ 1 \le j \le r : A(C_j,f) \cap D = \varnothing \}$ and $N = \{ 1 \le i \le \ell : A(R_i,f) \cap D = \varnothing \}$. We have $D \subset E_M$, and by Lemma 8.9 there exists $n \ge 1$ such that $D \subset G^n_N$; therefore $D \subset D^n_{N,M}$. On the other hand, if $j \notin M$ then by Lemma 8.3 $A(C_j,f) \subset D$, and if $i \notin N$ then by Lemma 8.9 $A(R_i,f) \subset D$. Hence

$$D^n_{N,M} - D \subset (G^n_N \cap E_M) - D \subset I - \bigcup_{j=1}^{r} A(C_j,f) \cup \bigcup_{i=1}^{\ell} A(R_i,f) ,$$

and so by Theorem 2.4 $\operatorname{int}(D^n_{N,M} - D) = \varnothing$. Proposition 8.5 thus gives us that $\mathbf{D}(f)$ is countable.

It remains to show that $\mathbf{D}(f)$ is finite if and only if each f-register-shift is tame. For $N \subset \{1,\ldots,\ell\}$ and $n \ge 1$ let $F^n_N = \kappa(G^n_N)$; we have $F^n_N \subset F^m_N$ for all $m \ge n$, and it is clear that $D^m_{N,M} = D^n_{N,M}$ whenever $F^m_N = F^n_N$. From this it follows that $\mathbf{D}(f)$ is finite if and only if there exists $p \ge 1$ such that $F^n_N = F^p_N$ for all $n \ge p$ and all $N \subset \{1,\ldots,\ell\}$. Suppose that each f-register-shift is tame; there thus exists $p \ge 1$ such that $A(K^i_p,f) - A(R_i,f)$ is countable for each $i = 1,\ldots, \ell$. Let $N \subset \{1,\ldots,\ell\}$ and $n \ge p$; we then have

$$G^n_N - G^p_N = \bigcup_{i\in N} A(K^i_p,f) - \bigcup_{i\in N} A(K^i_n,f) \subset \bigcup_{i\in N} (A(K^i_p,f) - A(R_i,f)) ,$$

which is countable, and hence $F^n_N = F^p_N$. Conversely, suppose there exists $p \ge 1$ such that $F^n_N = F^p_N$ for all $n \ge p$ and all $N \subset \{1,\ldots,\ell\}$. Let $1 \le i \le \ell$; then

$$A(K^i_p,f) - A(R_i,f) = A(K^i_p,f) - \bigcap_{n\ge p} A(K^i_n,f)$$

$$= \bigcup_{n\ge p} (A(K^i_p,f) - A(K^i_n,f)) = \bigcup_{n\ge p} (G^n_{\{i\}} - G^p_{\{i\}}) ,$$

which is countable, since $F^n_{\{i\}} = F^p_{\{i\}}$ for all $n \ge p$. Therefore R_i is tame. □

We now consider a generalization of Theorem 8.8 for mappings in $\mathbf{M}_0(I)$. Note that if (ψ,g) is a reduction of $f \in \mathbf{M}_0(I)$ then by Theorem 7.8 (6) $g \in \mathbf{M}_0(I)$. Let $f \in \mathbf{M}_0(I)$ and R be an f-register-shift; we say that an f-cycle K **supports** R if $R \subset K$ and $K - A(R,f)$ is "small", where by "small" we mean countable if R is tame and having no interior if R is not tame.

Remark: We can replace $K - A(R,f)$ by $A(K,f) - A(R,f)$ in this definition, since

$\text{int}(K - A(R,f)) = \emptyset$ (resp. $K - A(R,f)$ is countable) if and only if $\text{int}(A(K,f) - A(R,f)) = \emptyset$ (resp. $A(K,f) - A(R,f)$ is countable).

If R is an f-register-shift then clearly there exists an f-cycle K which supports R. Note that if K supports R then $A(K,f) \cap A(C,f) = \emptyset$ for each topologically transitive f-cycle C and $A(K,f) \cap A(R',f) = \emptyset$ for each f-register-shift $R' \neq R$.

Proposition 8.11 Let $f \in \mathbf{M}_0(I)$ and K be an f-cycle which supports some f-register-shift R; suppose that $I\text{-}A(K,f)$ is uncountable, and let (ψ,g) be a reduction of f with $\text{supp}(\psi) = \kappa(I\text{-}A(K,f))$. Then g has one register-shift less than f.

Proof Let $R, R_1,\ldots, R_\ell$ (with $\ell \geq 0$) be the f-register-shifts. For $i = 1,\ldots,\ell$ we have $R_i \cap A(K,f) = \emptyset$, and thus $R_i \cap \text{supp}(\psi) \neq \emptyset$; hence by Theorem 7.8 (4) $\psi(R_1),\ldots, \psi(_R\ell)$ are g-register-shifts. If there was another g-register-shift R' then by Proposition 7.12 we would have $R' = \psi(R)$. But this is not possible: Let $\{K_n\}_{n\geq 1}$ be a generator for R, so $R' \subset \psi(K_n)$ for each $n \geq 0$. Then by Lemma 2.15 $K_m \subset K$ for some $m \geq 0$, and hence $A(K_m,f) \subset I\text{-supp}(\psi)$, which implies that $\psi(K_m)$ is finite. Therefore g has one register-shift less than f. □

We call a reduction (ψ,g) of $f \in \mathbf{M}_0(I)$ a **register-shift** reduction if there exists an f-cycle K supporting some f-register-shift R such that $\text{supp}(\psi) = \kappa(I\text{-}A(K,f))$.

Proposition 8.12 Let $f \in \mathbf{M}_0(I)$ and $R_1,\ldots, R_\ell$ be the f-register-shifts. Then there exist $(\psi_1,f_1),\ldots, (\psi_\ell,f_\ell)$ such that (ψ_k,f_k) is a register-shift reduction of f_{k-1} for $k = 1,\ldots,\ell$ (with $f_0 = f$) if and only if $E = I - \bigcup_{i=1}^{\ell} A(R_i,f)$ is uncountable. Moreover, E will be uncountable if there is either a topologically transitive f-cycle or a non-tame f-register-shift.

Remark: If $(\psi_1,f_1),\ldots, (\psi_\ell,f_\ell)$ exist then by Proposition 8.11 we have $f_\ell \in \mathbf{M}_*(I)$.

Proof First note that we can always apply Proposition 8.11 $\ell-1$ times to obtain the register-shift reductions $(\psi_1,f_1),\ldots,(\psi_{\ell-1},f_{\ell-1})$. The problem lies with $f_{\ell-1}$, which has exactly one register-shift. Put $h = f_{\ell-1}$ and let R be the h-register-shift. If there exists a topologically transitive h-cycle C and K is an h-cycle supporting R then $A(C,h) \subset I-A(K,h)$, and so we can apply Proposition 8.11 to obtain the register-shift reduction (ψ_ℓ,f_ℓ) of $f_\ell-1$. Similarly, if R is not tame then we can find an h-cycle K supporting R such that $I-A(K,h)$ is uncountable; again we obtain the register-shift reduction (ψ_ℓ,f_ℓ) of $f_{\ell-1}$. Therefore the only case where we cannot apply Proposition 8.11 to h is when R is tame and $I-A(R,h)$ is countable. (If R is tame and $I-A(R,h)$ is uncountable then $I-A(K,h)$ is also uncountable for any h-cycle K supporting R .)

Suppose now there exists a topologically transitive f-cycle C ; then $A(C,f) \subset E$, and so E is uncountable. Also if (ψ,g) is a register-shift reduction of f then by Theorem 7.8 (3) $\psi(C)$ is a topologically transitive g-cycle. Thus in this case there is a topologically transitive h-cycle, and hence (ψ_ℓ,f_ℓ) exists. Suppose next that one of the f-register-shifts R_j is not tame, and let $\{K_n\}_{n\geq 1}$ be a generator for R_j . Then for n large enough we have $A(K_n,f)-A(R_j,f) \subset E$, and so again E is uncountable. Moreover, if K is an f-cycle supporting an f-register-shift $R_i \neq R_j$ and (ψ,g) is a reduction of f with $\mathrm{supp}(\psi) = \kappa(I-A(K,f))$ then $\psi(R_j)$ is a non-tame g-register-shift. Therefore in this case we can arrange that there is a non-tame h-register-shift, and hence (ψ_ℓ,f_ℓ) exists.

We are thus left with the case when there are no topologically transitive f-cycles and all the f-register-shifts are tame. For $i = 1,\ldots,\ell$ let K_i be an f-cycle supporting R_i ; then E is uncountable if and only if $I - \bigcup_{i=1}^{\ell} A(K_i,f)$ is uncountable, and from this it easily follows that if (ψ,g) is a reduction of f with $\mathrm{supp}(\psi) = \kappa(I-A(K_1,f))$ then E is uncountable if and only if $I - \bigcup_{i=2}^{\ell} A(\psi(R_i),g)$ is uncountable. Therefore $I-A(R,h)$ is uncountable if and only if E is uncountable, and hence also in this case (ψ_ℓ,f_ℓ) exists if and only if E is uncountable. □

Let $f \in M_o(I)$; as in the case of mappings from $M_*(I)$ we define a reduction

(ψ,g) of f to be **primary** if $\mathrm{supp}(\psi) = \kappa(I-A(C,f))$ for some topologically transitive f-cycle C.

Theorem 8.13 Let $f \in M_0(I)$ and (ψ,g) be a reduction of f with $\mathrm{supp}(\psi) \neq I$. Then there exist ℓ, $n \geq 0$ with $\ell + n \geq 1$ and $(\psi_1,f_1),\ldots,(\psi_{\ell+n},f_{\ell+n})$ such that (ψ_j,f_j) is a register-shift reduction of f_{j-1} for $j = 1,\ldots,\ell$ (with $f_0 = f$), (ψ_k,f_k) is a primary reduction of f_{k-1} for $k = \ell+1,\ldots,\ell+n$ and $(\psi,g) = (\psi_{\ell+n}\circ\cdots\circ\psi_1,f_{\ell+n})$.

Proof If R is an f-register-shift then by Lemma 8.9 either $R \subset \mathrm{supp}(\psi)$ or $R \cap \mathrm{supp}(\psi) = \varnothing$. Suppose R is an f-register-shift with $R \cap \mathrm{supp}(\psi) = \varnothing$; then again by Lemma 8.9 there exists an f-cycle K supporting R with $A(K,f) \cap \mathrm{supp}(\psi) = \varnothing$. Thus $\mathrm{supp}(\psi) \subset \kappa(I-A(K,f))$, and so there exists a reduction (ψ_1,f_1) of f with $\mathrm{supp}(\psi_1) = \kappa(I-A(K,f))$. By Theorem 7.4 there then also exists $\theta_1 \in V(I)$ such that $\psi = \theta_1\circ\psi_1$, since $\mathrm{supp}(\psi) \subset \mathrm{supp}(\psi_1)$. But by Proposition 8.11 f_1 has one less register-shift than f, and (θ_1,g) is a reduction of f_1. There therefore exists $\ell \geq 0$, $\theta_\ell \in V(I)$ and $(\psi_1,f_1),\ldots,(\psi_\ell,f_\ell)$ such that (ψ_j,f_j) is a register-shift reduction of f_{j-1} for $j = 1,\ldots,\ell$, $\psi = \theta_\ell\circ\psi_\ell\circ\cdots\circ\psi_1$ and so that $R \subset \mathrm{supp}(\theta_\ell)$ for each f_ℓ-register-shift R. Suppose θ_ℓ is not a homeomorphism; then by Lemma 8.9 and Theorem 2.4 there must be a topologically transitive f_ℓ-cycle C with $A(C,f_\ell) \cap \mathrm{supp}(\theta_\ell) = \varnothing$. There thus exists a primary reduction $(\psi_{\ell+1},f_{\ell+1})$ of f with $\mathrm{supp}(\psi_{\ell+1}) = \kappa(I-A(C,f_\ell))$, and since $\mathrm{supp}(\theta_\ell) \subset \mathrm{supp}(\psi_{\ell+1})$ there also exists $\theta_{\ell+1} \in V(I)$ with $\theta_\ell = \theta_{\ell+1}\circ\psi_{\ell+1}$. Moreover, we have $R \subset \mathrm{supp}(\theta_{\ell+1})$ for each $f_{\ell+1}$-register-shift R. (If R is an $f_{\ell+1}$-register-shift then by Proposition 7.12 $R = \psi_{\ell+1}(R')$ for some f_ℓ-register-shift R'. But $R' \subset \mathrm{supp}(\theta_\ell)$ and $\psi_{\ell+1}(\mathrm{supp}(\theta_\ell)) \subset \mathrm{supp}(\theta_{\ell+1})$.) Therefore if $\theta_{\ell+1}$ is not a homeomorphism then we can repeat this construction. Hence, so long as θ_ℓ, $\theta_{\ell+1},\ldots,\theta_{\ell+n-1}$ are not homeomorphisms, we obtain $(\psi_{\ell+1},f_{\ell+1}),\ldots,(\psi_{\ell+n},f_{\ell+n})$ and $\theta_{\ell+n} \in V(I)$ such that (ψ_k,f_k) is a primary reduction of f_{k-1} for $k = \ell+1,\ldots,\ell+n$ and $\psi = \theta_{\ell+n}\circ\psi_{\ell+n}\circ\cdots\circ\psi_1$. The proof will thus be complete if we can show that $\theta_{\ell+n}$ must be a homeomorphism for some $n \geq 0$. To show that this is the case consider the sets $D_p = \mathrm{supp}(\psi_{\ell+p}\circ\cdots\circ\psi_{\ell+1})$ for $p = 1,\ldots,n$. The sets $D_1,\ldots,D_n$ are all different, and if R is an f_ℓ-register-shift then it follows from the construction of ψ that $A(R,f_\ell) \subset D_p$

for each p . Hence $D_1, \ldots, D_n$ are n different elements from the set $\mathbf{D}'(f_\ell)$, where

$$\mathbf{D}'(f_\ell) = \{ D \in \mathbf{D}(f_\ell) : D \supset A(R,f) \text{ for each } f_\ell\text{-register-shift } R \} .$$

Let $D \in \mathbf{D}'(f_\ell)$, and put $M = \{ 1 \le j \le r : A(C_j,f_\ell) \cap D = \varnothing \}$, where $C_1, \ldots, C_r$ are the topologically transitive f_ℓ-cycles. Also let $E_M = I - \bigcup_{j \in M} A(C_j,f_\ell)$, and $D_M = \kappa(E_M)$. Then $D \subset D_M$ and $D_M \in \mathbf{D}(f)$. Moreover, if $j \notin M$ then by Lemma 8.3 $A(C_j,f_\ell) \subset D$, and so by Theorem 2.4 $\mathrm{int}(D_M - D) = \varnothing$. Therefore by Proposition 8.5 $\mathbf{D}'(f_\ell)$ is finite, and hence $\theta_{\ell+n}$ must be a homeomorphism for some $n \le |\mathbf{D}'(f_\ell)|$. □

We now apply some of the preceding results to show that if $f \in \mathbf{M}_0(I)$ has only one turning point then $\mathbf{D}(f)$ is linearly ordered by inclusion. Let $f \in \mathbf{M}_0(I)$ with $|T(f)| = 1$; then by Theorem 2.4 we have that exactly one of the following holds:

(8.1) There exists a single topologically transitive f-cycle C ; in this case $A(C,f)$ is dense in I .

(8.2) There exists a single f-register-shift R , and in this case $A(R,f)$ is dense in I .

If (8.1) holds then by Theorem 8.4 $\mathbf{D}(f)$ is finite; if (8.2) holds then by Theorem 8.10 $\mathbf{D}(f)$ is countable, and $\mathbf{D}(f)$ is finite if and only if R is tame.

Theorem 8.14 Let $f \in \mathbf{M}_0(I)$ with $|T(f)| = 1$; then $\mathbf{D}(f)$ is linearly ordered (by inclusion). Thus if $\mathbf{D}(f)$ is finite then we can write $\mathbf{D}(f) = \{D_1, \ldots, D_p\}$ so that $D_n \subset D_{n+1}$ for $n = 1, \ldots, p-1$. If $\mathbf{D}(f)$ is countably infinite then we can in fact write $\mathbf{D}(f) = \{D_1, D_2, \ldots\} \cup \{I\}$ so that $D_n \subset D_{n+1}$ for all $n \ge 1$.

Proof Suppose first that (8.1) holds. Then there is at most one primary reduction of f , because there is only one topologically transitive f-cycle. Let (ψ,g) be a reduction of f ; thus by Proposition 7.12 $g \in \mathbf{M}_*(I)$, and by Theorem 7.8 (1) $|T(g)| = 1$. (We cannot have $|T(g)| = 0$, since $g \in \mathbf{M}_0(I)$.) Hence (8.1) holds for g (i.e. there is a single topologically transitive g-cycle), and so in particular there is at most one primary reduction of g . Therefore, since $\mathbf{D}(f)$ is finite, there exists $m \ge 0$ and $(\psi_1,f_1), \ldots, (\psi_m,f_m)$ such that (ψ_k,f_k) is the unique primary reduction of f_{k-1} for $k = 1,\ldots, m$ (with $f_0 = f$) and such that there is no primary reduction of f_m . For $k = 1,\ldots, m$ let $D_k = \mathrm{supp}(\psi_k\circ\cdots\circ\psi_1)$; then $D_1 \supset D_2 \supset \cdots \supset D_m$ and by Theorems 7.4 and 8.8 $\mathbf{D}(f) - \{I\} = \{D_1,\ldots,D_m\}$. Hence $\mathbf{D}(f)$ is linearly ordered. Now suppose that (8.2) holds. For this case we need the following lemma:

Lemma 8.15 Let $f \in \mathbf{M}_0(I)$ with a single f-register-shift and no topologically transitive f-cycles; let (ψ,g) be a reduction of f with $\mathrm{supp}(\psi) \neq I$. Then $g \in \mathbf{M}_*(I)$.

Proof At the end of the proof of the theorem. □

Let (ψ,g) be a reduction of f with $\mathrm{supp}(\psi) \neq I$. Then by Lemma 8.15 we have $g \in \mathbf{M}_*(I)$, and thus in fact (8.1) holds for g , since we again have $|T(g)| = 1$. Therefore by the above $\mathbf{D}(g)$ is finite and linearly ordered, and hence by Proposition 7.11 $\{ D \in \mathbf{D}(f) : D \subset \mathrm{supp}(\psi) \}$ is finite and linearly ordered. This shows that $\{ D \in \mathbf{D}(f) : D \subset D' \}$ is finite and linearly ordered for each $D' \in \mathbf{D}(f)$ with $D' \neq I$. Let $\{K_n\}_{n\geq 1}$ be a generator for the single f-register-shift R , and for $n \geq 1$ put $D'_n = \kappa(I-A(K_n,f))$; then we have either $D'_n = \varnothing$ or $D'_n \in \mathbf{D}(f)$ for each $n \geq 1$. If $D \in \mathbf{D}(f)$ with $D \neq I$ then by Lemma 8.9 $D \subset I - A(K_n,f)$ for all large enough n , since by Theorem 2.4 $\overline{A(K_n,f)} = I$. It thus follows that $\mathbf{D}(f) - \{I\} = \bigcup_{n\geq 1} \{ D \in \mathbf{D}(f) : D \subset D'_n \}$. But $D'_n \neq I$ and $D'_n \subset D'_{n+1}$ for each $n \geq 1$, and so we can write $\mathbf{D}(f) - \{I\} = \{ D_n : 1 \leq n < N \}$, where $1 \leq N \leq \infty$ and $D_{n-1} \subset D_n$ for each $1 < n < N$. □

Proof of Lemma 8.15 Suppose there exists a g-register-shift R' and let $\{K'_n\}_{n\geq 1}$ be a generator for R' . For each $n \geq 1$ let $K_n = \psi^{-1}(K'_n)$ and put $R = \bigcap_{n\geq 1} K_n$; by Theorem 7.8 (5) $\{K_n\}_{n\geq 1}$ is a splitting sequence of f-cycles, and thus by Proposition 2.11 R is an f-register-shift, i.e. R is the single f-register-shift. Also the proof of Proposition 2.11 shows that $\{\hat{K}_n\}_{n\geq 1}$ is a generator for R , where $\hat{K}_n = \bigcap_{m\geq 0} f^m(K_n)$, and we have $\psi(\hat{K}_n) = K'_n$ for each n , because $\psi(f^m(K_n)) = g^m(\psi(K_n)) = g^m(K'_n) = K'_n$ and $\{f^m(K_n)\}_{m\geq 0}$ is a decreasing sequence of compact sets. But by Lemma 8.9 we must have $A(\hat{K}_n,f) \cap \mathrm{supp}(\psi) = \varnothing$ for all large enough n, since $\mathrm{supp}(\psi) \neq I$ and by Theorem 2.4 $\overline{A(\hat{K}_n,f)} = I$ for each $n \geq 1$. However, this is not possible, because it would imply that $K_n' = \psi(\hat{K}_n)$ is finite when n is large. Therefore there can be no g-register-shift, and hence $g \in \mathbf{M}_*(I)$. □

Let $f \in \mathbf{M}_0(I)$ with $|T(f)| = 1$ and let (ψ,g) be a reduction of f with $\mathrm{supp}(\psi) \neq I$. Then by Proposition 7.12 and Lemma 8.15 $g \in \mathbf{M}_*(I)$ and by Theorem 7.8 (1) $|T(g)| = 1$; thus (8.1) holds for g. The next result gives an additional property of g, which will play an important role in Section 12 (where we give a more detailed study of mappings with a single turning point).

Proposition 8.16 Let $f \in \mathbf{M}_0(I)$ with $|T(f)| = 1$ and let (ψ,g) be a reduction of f with $\mathrm{supp}(\psi) \neq I$. Then the single turning point of g is periodic.

Proof Since $\mathrm{supp}(\psi) \neq I$ we have by Lemmas 8.3 and 8.9 that there exists an f-cycle K with $\mathrm{supp}(\psi) \cap A(K,f) = \varnothing$. (If (8.1) holds for f then we can take K to be the single topologically transitive f-cycle; if (8.2) holds then a suitable element from a generator for the single f-register-shift will work.) Thus ψ maps each component of K onto a point, and hence $\psi(x) \in \mathrm{Per}(g)$ for each $x \in K$. In particular $\psi(\gamma)$ is periodic, where γ is the turning point of f (since $\gamma \in K$). But by Theorem 7.8 (1) $\psi(\gamma)$ is the turning point of g. □

We return now to the more general situation of mappings from $\mathbf{M}_0(I)$. Let $f \in \mathbf{M}_0(I)$; we end this section by analysing the contribution made to $\mathbf{D}(f)$ by each topologically transitive f-cycle and each f-register-shift.

Lemma 8.17 Let $f \in \mathbf{M}(I)$ and B be either a topologically transitive f-cycle or an f-register-shift; put $d(B,f) = \overline{A(B,f)}$. Then $d(B,f) \in \mathbf{D}(f)$, and so $d(B,f)$ is the smallest element of $\mathbf{D}(f)$ containing $A(B,f)$. Moreover, $d(B,f) = \overline{\mathrm{int}(d(B,f))}$.

Proof Let R be an f-register-shift and $\{K_n\}_{n\geq1}$ be a generator for R. Since $A(K_n,f)$ and $I - A(K_n,f)$ are both f-almost-invariant for each n, we have that $A(R,f)$ and $I - A(R,f)$ are also f-almost-invariant. Thus by Lemma 8.2 $\overline{A(R,f)}$ and

$I - \overline{A(R,f)}$ are f-almost-invariant. Moreover, if U is any open subset of I then $\overline{U}$ is perfect, and so $\overline{A(R,f)} = \kappa(\overline{A(R,f)})$; hence by Lemma 7.6 $d(R,f) \in \mathbf{D}(f)$. Now by Lemma 8.9 we have that $A(K_n,f) \subset d(R,f)$ for all large enough n , and so in particular $\text{int}(d(R,f)) \neq \varnothing$. Put $D = \overline{\text{int}(d(R,f))}$; then $D \neq \varnothing$, and thus by Lemmas 7.6 and 8.2 $D \in \mathbf{D}(f)$. But $A(R,f) \subset D$ (because $A(K_n,f) \subset D$ for all large enough n) and $D \subset d(R,f)$; hence $D = d(R,f)$, i.e. we have $d(R,f) = \overline{\text{int}(d(R,f))}$. The proof for a topologically transitive f-cycle is similar.
□

Let $f \in \mathbf{M}(I)$; if B is either a topologically transitive f-cycle or an f-register-shift then we put

$$\mathbf{D}(f,B) = \{ D \in \mathbf{D}(f) : D \subset d(B,f) \} .$$

The next result shows that if $f \in \mathbf{M}_0(I)$ then each element $D \in \mathbf{D}(f)$ can be written in the form $D = D_1 \cup \cdots \cup D_m$, where $D_j \in \mathbf{D}(f,B_j)$ for $j = 1,\dots, m$ and each B_j is either a topologically transitive f-cycle or an f-register-shift.

Proposition 8.18 Let $f \in \mathbf{M}_0(I)$ with topologically transitive f-cycles $C_1,\dots, C_r$ and f-register-shifts $R_1,\dots, R_\ell$; let $D \in \mathbf{D}(f)$ and for $1 \le j \le r$, $1 \le i \le \ell$ put $E_j = \kappa(D \cap d(C_j,f))$ and $F_i = \kappa(D \cap d(R_i,f))$; also let $M = \{ 1 \le j \le r : E_j \neq \varnothing \}$ and $N = \{ 1 \le i \le \ell : F_i \neq \varnothing \}$. Then $E_j \in \mathbf{D}(f,C_j)$ for each $j \in M$, $F_i \in \mathbf{D}(f,R_i)$ for each $i \in N$ and $D = \bigcup_{j\in M} E_j \cup \bigcup_{i\in N} F_i$.

Proof It is clear that for each j either $E_j = \varnothing$ or $E_j \in \mathbf{D}(f,C_j)$ and that for each i either $F_i = \varnothing$ or $F_i \in \mathbf{D}(f,R_i)$. Now put $D' = \bigcup_{j\in M} E_j \cup \bigcup_{i\in N} F_i$; thus $D' = E_1 \cup \cdots \cup E_r \cup F_1 \cup \cdots \cup F_\ell$. By Theorem 2.4 we have that $A(C_1,f) \cup \cdots \cup A(C_r,f) \cup A(R_1,f) \cup \cdots \cup A(R_\ell,f)$ is dense in I and so

$$d(C_1,f) \cup \cdots \cup d(C_r,f) \cup d(R_1,f) \cup \cdots \cup d(R_\ell,f) = I .$$

Hence $D - D'$ is countable, and therefore $D = D'$, because D is perfect. □

Let $f \in \mathbf{M}_0(I)$ and B be either a topologically transitive f-cycle or an f-register-shift; then there exists a reduction (ψ,g) of f with $\text{supp}(\psi) = d(B,f)$, and by Proposition 7.11 $\hat{\psi}$ maps $\mathbf{D}(g)$ bijectively onto

$\mathbf{D}(f,B)$. Thus the structure of $\mathbf{D}(f,B)$ can be determined from the structure of $\mathbf{D}(g)$. We now see which mappings g can occur when (ψ,g) is a reduction of f with $\mathrm{supp}(\psi) = d(B,f)$.

Let $\mathbf{M}_t(I)$ (resp. $\mathbf{M}_r(I)$) denote the set of mappings $g \in \mathbf{M}_0(I)$ having exactly one topologically transitive g-cycle and no g-register-shifts (resp. having exactly one g-register-shift and no topologically transitive g-cycles).

Proposition 8.19 Let $f \in \mathbf{M}_0(I)$ and (ψ,g) be a reduction of f ; then $g \in \mathbf{M}_0(I)$. Moreover:

(1) If $\mathrm{supp}(\psi) = d(C,f)$ for some topologically transitive f-cycle C then $g \in \mathbf{M}_r(I)$.

(2) If $\mathrm{supp}(\psi) = d(R,f)$ for some f-register-shift R then $g \in \mathbf{M}_t(I)$.

Proof We have already noted that $g \in \mathbf{M}_0(I)$.

(1): Let $K = \psi(C)$; then by Theorem 7.8 (3) K is a topologically transitive g-cycle. Let K' be any g-cycle and put $C' = \psi^{-1}(K')$; then C' is an f-cycle and $\mathrm{int}(C) \cap \mathrm{int}(C') \neq \varnothing$ (since if $\mathrm{int}(C) \cap \mathrm{int}(C') = \varnothing$ then $A(C,f) \cap \mathrm{int}(C') = \varnothing$ and so $\mathrm{supp}(\psi) \cap \mathrm{int}(C') = \varnothing$, and this would imply that $K' = \psi(C')$ is finite). Thus $C \cap C'$ is a closed f-invariant subset of C with $\mathrm{int}(C \cap C') \neq \varnothing$, and hence $C \cap C' = C$, i.e. $C \subset C'$. Therefore $K' = \psi(C') \supset \psi(C) = K$, and so by Proposition 2.2 and Lemma 2.15 there is no other topologically transitive g-cycle and there are no g-register-shifts.

(2): By Theorem 7.8 (4) $\psi(R)$ is a g-register-shift. Let L be any g-cycle and put $C = \psi^{-1}(L)$; then C is an f-cycle, and if $\{K_n\}_{n\geq 1}$ is a generator for R then, as in (1), $\mathrm{int}(C) \cap \mathrm{int}(K_n) \neq \varnothing$ for each n . Thus by Lemma 2.15 $K_n \subset C$ for all large enough n , and hence $\psi(R) \subset \psi(K_n) \subset \psi(C) = L$. Therefore by Proposition 2.2 and Lemma 2.15 there is no other g-register-shift and there are no topologically transitive g-cycles. □

Proposition 8.20 Let $f \in \mathbf{M}_0(I)$. Then $\mathbf{D}(f,C)$ is finite for each topologically transitive f-cycle C , $\mathbf{D}(f,R)$ is countable for each f-register-shift R and $\mathbf{D}(f,R)$ is finite if and only if R is tame.

Proof This follows immediately from Theorem 8.10, Proposition 8.19 and Proposition 7.11. □

Let $h \in M_t(I)$ with $|T(h)| = 1$; then by Theorem 8.14 $D(h)$ is linearly ordered. The simple stucture of $D(h)$ in this case is really a consequence of the fact that if $f \in M_t(I)$ with $|T(f)| = 1$ and (ψ,g) is a reduction of f then also $g \in M_t(I)$. For a general mapping $f \in M_t(I)$ this is no longer true, as can be seen from the example at the end of Section 7. In this example we had a mapping $f \in M_t(I)$ and a reduction (ψ,g) of f such that there are two topologically transitive g-cycles; thus $g \notin M_t(I)$. However, we now show that this example represents the worst that can happen. Let $M_s(I)$ denote the set of mappings $f \in M_*(I)$ having either one topologically transitive f-cycle C or two topologically transitive f-cycles C_1 and C_2 with $C_1 \cap C_2 \neq \emptyset$. (Note that if C_1 and C_2 are two (different) topologically transitive f-cycles then $C_1 \cap C_2 \subset \partial C_1 \cap \partial C_2$, since $\mathrm{int}(C_1) \cap \mathrm{int}(C_2) = \emptyset$.)

Proposition 8.21 (1) If (ψ,g) is a reduction of $f \in M_t(I)$ then $g \in M_s(I)$.
(2) If (ψ,g) is a reduction of $f \in M_r(I)$ with $\mathrm{supp}(\psi) \neq I$ then $g \in M_s(I)$.

Proof (1): By Proposition 7.12 we have $g \in M_*(I)$. Let C be the single topologically transitive f-cycle and let K' be any g-cycle. Then $K = \psi^{-1}(K')$ is an f-cycle and $\mathrm{int}(K) \cap A(C,f) \neq \emptyset$, since $A(C,f)$ is dense in I . Thus also $\mathrm{int}(K) \cap \mathrm{int}(C) \neq \emptyset$, and so $C \cap K$ is a closed f-invariant subset of C with $\mathrm{int}(C \cap K) \neq \emptyset$; hence $C \cap K = C$, i.e. $C \subset K$. Therefore $K' = \psi(K) \supset \psi(C)$, and in particular we have $C' \supset \psi(C)$ for each topologically transitive g-cycle C' . From this it immediately follows that $g \in M_s(I)$.

(2): By Lemma 8.15 we have $g \in M_*(I)$. Let $\{K_n\}_{n\geq 1}$ be a generator for the single f-register-shift R , and let C' be any g-cycle. Then $C = \psi^{-1}(C')$ is an f-cycle and for each $n \geq 1$ we have $\mathrm{int}(C) \cap \mathrm{int}(K_n) \neq \emptyset$, since $A(K_n,f)$ is dense in I . Thus by Lemma 2.15 $K_n \subset C$ for all large enough n , and hence $C' = \psi(C) \supset \psi(K_n)$ for all large enough n . As in (1), this implies that $g \in M_s(I)$. □

It is not difficult to construct mappings $f \in \mathbf{M}_r(I)$ for which $\mathbf{D}(f)$ is not linearly ordered. However, we now show that if $f \in \mathbf{M}_r(I)$ then it is possible to remove finitely many elements from $\mathbf{D}(f)$ so that what remains is linearly ordered. For this we first need a slight generalization of Theorem 8.8.

Proposition 8.22 Let $f \in \mathbf{M}_0(I)$ and (ψ,g), (ψ',g') be reductions of f with $g \in \mathbf{M}_*(I)$ and $\mathrm{supp}(\psi') \subsetneq \mathrm{supp}(\psi)$. Then there exists $n \geq 1$ and $(\psi_1,g_1),\ldots,(\psi_n,g_n)$ such that (ψ_k,g_k) is a primary reduction of g_{k-1} for $k = 1,\ldots,n$ (with $g_0 = g$) and $(\psi',g') = (\psi_n\circ\cdots\circ\psi_1\circ\psi, g_n)$.

Proof By Proposition 7.1 (3) there exists $\theta \in \mathbf{V}(I)$ with $\psi' = \theta\circ\psi$; thus (θ,g') is a reduction of g, and $\mathrm{supp}(\theta) \neq I$ because $\mathrm{supp}(\psi') \neq \mathrm{supp}(\psi)$. Therefore by Theorem 8.8 there exists $n \geq 1$ and $(\psi_1,g_1),\ldots,(\psi_n,g_n)$ such that (ψ_k,g_k) is a primary reduction of g_{k-1} for $k = 1,\ldots,n$ (with $g_0 = g$) and $(\theta,g') = (\psi_n\circ\cdots\circ\psi_1,g_n)$. Hence also $\psi' = \theta\circ\psi = \psi_n\circ\cdots\circ\psi_1\circ\psi$. □

Proposition 8.23 Let $f \in \mathbf{M}_r(I)$ with $\mathbf{D}(f)$ infinite (i.e. the single f-register-shift is not tame). Then we can write $\mathbf{D}(f) = \mathbf{D}_0 \cup \{D_1,D_2,\ldots\} \cup \{I\}$ with $\mathbf{D}_0$ finite, $D \subset D_1$ for all $D \in \mathbf{D}_0$ and $D_n \subset D_{n+1}$ for each $n \geq 1$.

Proof Let R be the single f-register-shift and $\{K_n\}_{n\geq1}$ a generator for R; for $n \geq 1$ let $D'_n = \kappa(I-A(K_n,f))$. Then, as in the proof of Theorem 8.14, we have for each $n \geq 1$ that either $D'_n = \varnothing$ or $D'_n \in \mathbf{D}(f)$, and that $D'_n \subset D'_{n+1}$; also $\mathbf{D}(f)-\{I\} = \bigcup_{n\geq1} \{ D \in \mathbf{D}(f) : D \subset D'_n \}$. Now since $\mathbf{D}(f)$ is infinite we can assume without loss of generality that $D'_1 \neq \varnothing$ and $D'_n \neq D'_{n+1}$ for all $n \geq 1$. For $n \geq 1$ let (ψ_n,f_n) be a reduction of f with $\mathrm{supp}(\psi_n) = D'_n$. Then by Theorem 7.8 (1) we have $|T(f_n)| \leq |T(f_{n+1})| \leq |T(f)|$ for all $n \geq 1$ (since by Proposition 7.1 (3) there exists $\theta_n \in \mathbf{V}(I)$ with $\psi_n = \theta_n\circ\psi_{n+1}$); thus let $\alpha = \lim_{n\to\infty} |T(f_n)|$ and $q \geq 1$ be such that $|T(f_n)| = \alpha$ for all $n \geq q$.

Lemma 8.24 Let (φ,h) be a reduction of f with $\mathrm{supp}(\varphi) \neq I$ (and so $h \in \mathbf{M}_s(I)$). If $|T(h)| = \alpha$ then there is only one topologically transitive h-cycle.

Proof At the end of the proof of the proposition. □

Now fix $p > q$ and let $D \in \mathbf{D}(f)$ with $D \subsetneq D'_p$; let (ψ,g) be a reduction of f with $\text{supp}(\psi) = D$. By Lemma 8.15 $f_p \in \mathbf{M}_*(I)$ and so by Proposition 8.22 there exists $m \geq 1$ and $(\varphi_1,g_1),\ldots,(\varphi_m,g_m)$ such that (φ_k,g_k) is a primary reduction of g_{k-1} for $k = 1,\ldots,m$ (with $g_0 = f_p$) and $(\psi,g) = (\varphi_m\circ\cdots\circ\varphi_1\circ\psi_p,g_m)$. For $k = 1,\ldots,m$ let $D''_k = \text{supp}(\varphi_k\circ\cdots\circ\varphi_1\circ\psi_p)$; then $D'_p \supset D''_1 \supset \cdots \supset D''_m = D$. Let $1 \leq \ell \leq m$; if $D''_\ell \supset D'_q$ then

$$\alpha = |T(f_p)| \geq |T(g_1)| \geq \cdots \geq |T(g\ell)| \geq |T(f_q)| = \alpha ,$$

and so by Lemma 8.24 (φ_k,g_k) is the only primary reduction of g_{k-1} for $k = 1,\ldots,\ell$. Thus if $D'_q \not\subset D$ then we must have $D'_q = D''_\ell$ for some $1 \leq \ell \leq m$, and so in particular $D \subset D'_q$. This shows that

$$\{ D \in \mathbf{D}(f) : D \subset D'_p \} = \{ D \in \mathbf{D}(f) : D \subset D'_q \} \cup \{ D \in \mathbf{D}(f) : D'_q \subset D \subset D'_p \} ,$$

and that $\{ D \in \mathbf{D}(f) : D'_q \subset D \subset D'_p \}$ is linearly ordered. The result therefore follows with $D_0 = \{ D \in \mathbf{D}(f) : D \subset D'_q \}$. □

Proof of Lemma 8.24 Suppose there are two topologically transitive h-cycles C_1 and C_2 ; then by the proof of Proposition 8.21 (2) there exists $n \geq q$ such that $\text{supp}(\varphi) \subset D'_{n+1}$ and $\varphi(K_n) \subset C_1 \cap C_2$. Now $\text{int}(K_n) \cap \text{supp}(\psi_{n+1}) \neq \emptyset$ (since otherwise $\text{supp}(\psi_{n+1}) \subset I\text{-}A(K_n,f)$, and this would imply that $D'_{n+1} = \text{supp}(\psi_{n+1}) \subset \kappa(I\text{-}A(K_n,f)) = D'_n$), and thus by Theorem 7.8 (2) $C = \psi_{n+1}(K_n)$ is an f_{n+1}-cycle. By Proposition 7.3 (1) we can find $\theta \in V(I)$ so that $\varphi = \theta\circ\psi_{n+1}$, and we then have $\theta(C) = \varphi(K_n) \subset C_1 \cap C_2 = \partial C_1 \cap \partial C_2$. But $(C_1 \cap C_2) \cap T(h) = \emptyset$ (otherwise $\text{int}(C_1) \cap \text{int}(C_2) \neq \emptyset$), and $\text{int}(C) \cap T(f_{n+1}) \neq \emptyset$ (since $f_{n+1} \in \mathbf{M}_0(I)$), and hence by Theorem 7.8 (1) $|T(f_{n+1})| > |T(h)|$. However, this contradicts the fact that $|T(f_{n+1})| = \alpha = |T(h)|$, and therefore there is only one topologically transitive h-cycle. □

9. COUNTABLE CLOSED INVARIANT SETS

For $f \in \mathbf{M}(I)$ let $\mathbf{I}(f)$ denote the set of closed subsets D of I such that D is f-invariant and $I-D$ is f-almost-invariant (so $D(f) \subset \mathbf{I}(f)$). In this section we study the countable elements in $\mathbf{I}(f)$. Note that if $f \in \mathbf{M}(I)$ and $I-Z(f)$ is countable then $I-Z_*(I)$ is also countable, and by Lemma 2.1 $I-Z_*(I) \in \mathbf{I}(f)$. **Theorem 9.5** deals with this case, and shows that if $I-Z(f)$ is countable then there exists $p \geq 0$ such that for each $x \in I$ the sequence $\{f^{2^p n}(x)\}_{n\geq 1}$ converges to a periodic point of f ; in particular this implies that each periodic point of f has a period which divides 2^p .

We also obtain a similar result for the case when we have f-cycles C' and C with $C' \subset C$ and $C - A(C',f)$ countable. Theorem 9.6 shows that there then exist p , $q \geq 0$ such that $\mathrm{per}(C') = 2^p\,\mathrm{per}(C)$ and each periodic point in $C - A(C',f)$ has a period which divides $2^q\,\mathrm{per}(C)$. Moreover, each point in $C - A(C',f)$ is eventually periodic. ($x \in I$ is eventually periodic if $f^n(x) \in \mathrm{Per}(f)$ for some $n \geq 0$.) This result is then used to analyse the structure of tame register-shifts. Let R be an f-register-shift and K be an f-cycle with $R \subset K$ such that $K - A(R,f)$ is countable (i.e. R is tame). Theorem 9.7 then shows that

$$K - A(R,f) \;=\; \{\, x \in K : \; x \text{ is eventually periodic } \},$$

and if $x \in K \cap \mathrm{Per}(f)$ then x has period $2^j\,\mathrm{per}(K)$ for some $j \geq 0$. Moreover, if $\{K_n\}_{n\geq 1}$ is a generator for R with $K_1 \subset K$ then there exists a strictly increasing sequence $\{q_n\}_{n\geq 1}$ of non-negative integers such that $\mathrm{per}(K_n) = 2^{q_n}\,\mathrm{per}(K)$ for all $n \geq 1$.

Let $f \in \mathbf{M}(I)$ and $D \in \mathbf{I}(f)$; by Lemma 8.1 we have $D \in \mathbf{I}(f^n)$ for all $n \geq 1$. The next result is a variation on results in Block (1977), (1979) and Misiurewicz (1980); (see also Section 1 in Nitecki (1982)).

Proposition 9.1 Let $f \in \mathbf{M}(I)$ and $D \in \mathbf{I}(f)$ be countable. Then there exists $q \geq 0$ such that for each $x \in D$ the sequence $\{f^{2^q n}(x)\}_{n\geq 1}$ converges to a periodic point of f in D ; in particular, each periodic point of f in D has a period which divides 2^q .

Proof The following lemma is crucial. It is a weak form of an idea which occurs in Block (1977), (1979), Misiurewicz (1980), and in the proof of Šarkovskii's theorem (Šarkovskii (1964), Štefan (1977)) given in Block, Guckenheimer, Misiurewicz and Young (1980).

Lemma 9.2 Let $f \in \mathbf{M}(I)$ and $D \in \mathbf{I}(f)$ be countable; if $x \in D$ and $x < f(x) < f^2(x)$ then $x < f^n(x)$ for all $n \geq 1$. (Similarly, if $x \in D$ and $x > f(x) > f^2(x)$ then $x > f^n(x)$ for all $n \geq 1$.)

Proof It simplifies the proof to assume that $I-D$ is actually f-invariant (rather than just f-almost-invariant), and we can always reduce things to this case: If $x \in I-D$ with $f(x) \in D$ then $x \in S(f)$ and we can alter f in a small neighbourhood of x (without changing f on D) to ensure that $f(x) \in I-D$. Thus let $D \subset I$ be closed with both D and $I-D$ f-invariant, and let $x \in D$ with $x < f(x) < f^2(x)$. Suppose $x \geq f^n(x)$ for some $n \geq 3$; we show that this implies D is uncountable. By replacing x with $f^{k-2}(x)$, where $k = \max\{ j \geq 2 : x < f(x) < \cdots < f^j(x) \}$, we can assume that $f^3(x) < f^2(x)$. Put $y = f(x)$ and $z = f^2(x)$. Since $f(y) > y$ and $f(z) < z$, we can find $v \in (y,z)$ with $f(v) = v$; also, since $f(x) = y < v$ and $f(y) = z > v$, there exists $u \in (x,y)$ with $f(u) = v$. Consider $g = f^{n-1}$ on $[u,v]$; we have $g(v) = v$, $g(u) = v$ and $g(y) = f^n(x) \leq x < u$. Let β be the smallest element in $(y,v] \cap \mathrm{Fix}(g)$; then we can find $\alpha \in [u,y)$ with $g(\alpha) = \beta$, and we have $g(y) < \alpha$. (See the picture on the following page.)

Suppose $\beta \in I-D$, and let U be the component of $I-D$ containing β. Then $g(\bar{U}) \subset \bar{U}$, because $I-D$ is g-invariant and $\beta \in \mathrm{Fix}(g)$. Thus if $\bar{U} = [c,d]$ then $g(c) \geq c$, and hence $y \in U$. However, this is not possible, since $y = f(x) \in D$. Therefore $\beta \in D$, and so, again from the fact that $I-D$ is g-invariant, we have $w \in D$ whenever $g^k(w) = \beta$ for some $k \geq 0$. But it is not difficult to use this to show that $[\alpha,\beta] \cap D$ contains a Cantor-like set; (for more details see Lemma 13.7). Thus D is uncountable. □

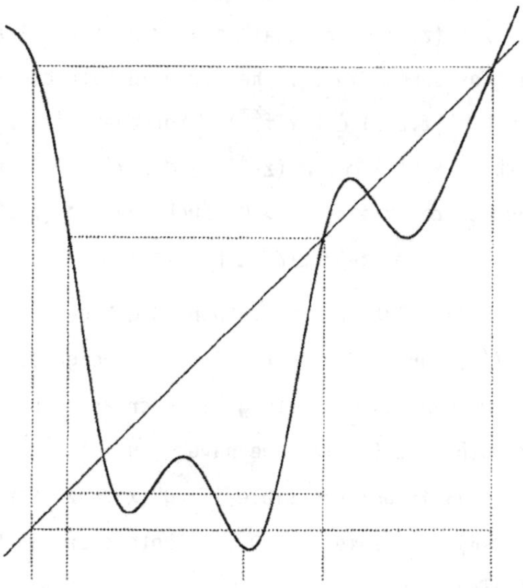

Lemma 9.3 Let $D \in \mathbf{I}(f)$ be countable; then we have:

(1) Each periodic point of f in D has period 2^n for some n .

(2) If $z \in Per(m,f) \cap D$ then there exists a neighbourhood U of z such that $Per(f) \cap D \cap U \subset Fix(f^{2m})$.

(3) If $x , z \in D$ with $z \in Fix(f)$ and there exists a subsequence $\{n_k\}_{k\geq 1}$ such that $\lim_{k\to\infty} f^{n_k}(x) = z$ then in fact $\lim_{n\to\infty} f^n(x) = z$.

Proof (1): Let $x \in Per(m,f) \cap D$ with $m > 1$; then by Lemma 9.2 $f^k(x) - f^{k+1}(x)$ and $f^{k+1}(x) - f^{k+2}(x)$ have opposite signs for each $k \geq 0$, and hence m must be even, say $m = 2\ell$. But $D \in \mathbf{I}(f^2)$ and $x \in Per(\ell,f^2) \cap D$; thus if $\ell > 1$ then the same argument shows that ℓ is even. Repeating this gives us that $m = 2^n$ for some $n \geq 1$.

(2): Put $g = f^{2m}$, so $D \in I(g)$ and $z \in Fix(g)$. We have either $z \in T(g)$ or g is increasing in a neighbourhood of z , and thus without loss of generality we can assume that g is increasing on $[z,z+\varepsilon]$ for some $\varepsilon > 0$. Now if $g(x) > x$ for all $x \in (z,y)$ for some $y \in (z,z+\varepsilon]$ then, applying Lemma 9.2 to g, we have that $\mathrm{Per}(g) \cap D \cap (z,z+\delta) = \emptyset$ for some $\delta > 0$. On the other hand, if for each $y \in (z,z+\varepsilon]$ there exists $x \in (z,y)$ such that $g(x) \le x$ then clearly $\mathrm{Per}(g) \cap (z,z+\delta) \subset \mathrm{Fix}(g)$ for some $\delta > 0$. Therefore in both cases there exists $\delta > 0$ such that $\mathrm{Per}(f) \cap D \cap [z,z+\delta) \subset \mathrm{Fix}(f^{2m})$. Moreover, if $z \notin T(g)$ then the same argument also shows that $\mathrm{Per}(f) \cap D \cap (z-\delta',z] \subset \mathrm{Fix}(f^{2m})$ for some $\delta' > 0$. Finally, if $z \in T(g)$ then there exists $\delta'' > 0$ such that $g((z-\delta'',z]) \subset [z,z+\delta)$, and this implies that $\mathrm{Per}(f) \cap D \cap (z-\delta'',z+\delta) \subset \mathrm{Fix}(f^{2m})$.

(3): This is very similar to (2). Put $g = f^2$; then $D \in \mathbf{I}(g)$, $z \in \mathrm{Fix}(g)$, and again we have either $z \in T(g)$ or g is increasing in a neighbourhood of z. Thus without loss of generality we can assume that g is increasing on $[z,z+\varepsilon]$ for some $\varepsilon > 0$ and that for each $y \in (z,z+\varepsilon)$ we have $\{ n \ge 1 : g^n(x) \in [z,y] \}$ is infinite. But by Lemma 9.2 this is only possible if $g^n(x) = z$ for some $n \ge 0$ or $g(y) < y$ for all $y \in (z,z+\delta)$ for some $\delta > 0$. In both cases we have $\lim_{n\to\infty} g^n(x) = z$, and hence also $\lim_{n\to\infty} f^n(x) = z$. □

We now need the following well-known fact:

Lemma 9.4 Let $g \in \mathbf{C}(I)$ and A be a non-empty closed subset of I with $g(A) \subset A$. If A is countable then $\mathrm{Per}(g) \cap A \ne \emptyset$.

Proof For $x \in A$ let $E_x = \{ g^n(x) : n \ge 0 \}$ and put $B_x = \overline{E_x}$; thus $B_x \subset A$, $g(B_x) \subset B_x$ and $B_y \subset B_x$ for each $y \in B_x$. We now introduce a partial order $\le$ on A by defining $x \le y$ if $B_y \subset B_x$. Then each totally ordered subset M of A has an upper bound: Since A is compact $\bigcap_{x\in M} B_x$ is non-empty, and if $z \in \bigcap_{x\in M} B_x$ then $x \le z$ for all $x \in M$. Thus by Zorn's lemma there exists a maximal element $w \in A$ with respect to $\le$. In particular we then have $B_x = B_w$ for all $x \in B_w$, and hence $x \in B_{g(x)}$ for each $x \in B_w$ (because $x \in B_x = B_w = B_{g(x)}$); i.e. $x \in \overline{\{ g^n(x) : n \ge 1 \}}$ for each $x \in B_w$. But B_w is a countable closed set, and so by the Baire category theorem B_w contains an isolated point v (see Lemma 13.2); and we must then have $v \in \{ g^n(v) : n \ge 1 \}$, i.e. $v \in \mathrm{Per}(g)$. □

It follows from Lemmas 9.3 (2) and 9.4 that if $D \in \mathbf{I}(f)$ is countable then

$\{ n \geq 1 : Per(n,f) \cap D \neq \emptyset \}$ is finite. To see this, suppose it were not the case, and let A be the set of all points $x \in I$ for which there exists a sequence $\{x_n\}_{n\geq 1}$ with $x_n \in Per(m_n,f) \cap D$ and $\lim_{n\to\infty} m_n = \infty$ such that $x = \lim_{n\to\infty} x_n$. Then A is a non-empty, closed subset of I and $f(A) \subset A$; moreover, A is countable, since $A \subset D$. Therefore by Lemma 9.4 $Per(f) \cap A \neq \emptyset$. But by Lemma 9.3 (2) this is clearly not possible.

Let $D \in \mathbf{I}(f)$ be countable; using the above together with Lemma 9.3 (1) we have now shown that there exists $q \geq 0$ such that each periodic point of f in D has a period which divides 2^q. It remains to show that the sequence $\{f^{2^q n}(x)\}_{n\geq 0}$ converges for each $x \in D$. Put $g = f^{2^q}$ and let A be the set of accumulation points of the sequence $\{g^n(x)\}_{n\geq 0}$ (i.e. $z \in A$ if there exists a subsequence $\{n_k\}_{k\geq 1}$ with $z = \lim_{k\to\infty} g^{n_k}(x)$.) Then A is a non-empty closed subset of D with $g(A) \subset A$, and hence by Lemma 9.4 $Per(g) \cap A \neq \emptyset$. But $Per(g) \cap A \subset Fix(g)$, and thus there exists a subsequence $\{n_k\}_{k\geq 1}$ and $z \in Fix(g)$ such that $z = \lim_{k\to\infty} g^{n_k}(x)$. Now by Lemma 9.3 (3) (applied to g) we have $\lim_{n\to\infty} g^n(x) = z$, i.e. the sequence $\{f^{2^q n}(x)\}_{n\geq 0}$ converges. This completes the proof of Proposition 9.1.
□

Let $f \in \mathbf{M}(I)$; we call $x \in I$ **eventually periodic** if $f^n(x) \in Per(f)$ for some $n \geq 0$.

Theorem 9.5 Let $f \in \mathbf{M}(I)$ with $I\text{-}Z(f)$ countable. Then we have:

(1) There exists $p \geq 0$ such that for each $x \in I$ the sequence $\{f^{2^p n}(x)\}_{n\geq 0}$ converges to a periodic point of f; in particular, each periodic point of f has a period which divides 2^p.

(2) $Homt(f) = \emptyset$.

(3) $Per(f) \cap (I\text{-}Z_*(f))$ is finite, and each point in $I\text{-}Z_*(f)$ is eventually periodic.

(4) Each component of $Z_*(f)$ is eventually periodic.

Proof Put $D = I\text{-}Z_*(f)$, so $D \in \mathbf{I}(f)$. Let $q \geq 0$ be as in Proposition 9.1, and let $g = f^{2^q}$. Then $Z_*(g) = Z_*(f)$ and hence $D = I-Z_*(g)$.

(3): We have $Per(g) = Per(f)$, and by Proposition 9.1 $Per(g) \cap D = Fix(g) \cap D$. If $u, v \in Fix(g) \cap D$ with $u < v$ and $(u,v) \cap T(g) = \emptyset$ then $(u,v) \subset Sink(g)$, and so $(u,v) \cap D = \emptyset$. Thus $Fix(g) \cap D$ is finite, i.e. $Per(f) \cap (I-Z_*(f))$ is finite. Let $x \in D$; then by Proposition 9.1 $\lim_{n\to\infty} g^n(x) = z$ for some $z \in Fix(g) \cap D$, and hence we must have $g^n(x) = z$ for some $n \geq 0$. (Let $h \in M(I)$ and $\lim_{n\to\infty} h^n(y) = w$ for some $w \in Fix(h)$; if $h^n(y) \neq w$ for all $n \geq 0$ then there exists a sink J of h and $m \geq 0$ such that $h^n(y) \in J$ for all $n \geq m$.) Therefore each point in $I-Z_*(f)$ is eventually periodic.

(4): This follows from (3), since if U is a component of $Z_*(f)$ then $\partial U \subset D$ and each point of D is eventually periodic.

(1): By Proposition 9.1 $\{g^n(x)\}_{n\geq 0}$ converges to a periodic point of f for each $x \in D$; thus, by (4), we need only consider points $x \in I$ lying in some periodic component of $I-D$. Let U be a periodic component of $I-D$ with period m . Note that if $U \neq [a,b]$ then $\bar{U} \cap Per(f) \cap D \neq \emptyset$, and so $m | 2^{q+1}$. If $U \cap Fix(f^m) = \emptyset$ then clearly $\lim_{n\to\infty} f^{mn}(x) = w$ for all $x \in U$, where w is one of the end-points of $\bar{U}$. Assume then that $U \cap Fix(f^m) \neq \emptyset$ and let $z \in U \cap Fix(f^m)$; also let $B \subset U$ be a closed interval containing z . For each $x \in B$ there exists $j \geq 0$ and $\varepsilon > 0$ such that $f^j((x-\varepsilon, x+\varepsilon)) \subset Z(f)$, and hence by Proposition 4.1 we have $f^k((x-\varepsilon, x+\varepsilon)) \subset Z(f)$ for all $k \geq j$. Since B is compact there thus exists $n \geq 0$ such that $f^k(B) \subset Z(f)$ for all $k \geq n$. Choose $\ell \geq 0$ so that $\ell m \geq n$, and let J be the component of $Z(f)$ containing $f^{\ell m}(B)$; then $z \in J$ (because $z \in f^{\ell m}(B)$), and therefore J must be the maximal sink of f containing z . Since J is defined independently of B it follows that for each $x \in U$ there exists $k \geq 0$ such that $f^k(x) \in J$. But f^m is monotone on J and $f^m(J) \subset J$, and hence for each $x \in U$ the sequence $\{f^{2mn}(x)\}_{n\geq 0}$ converges to a periodic point of f in $\bar{J}$. This shows that (1) holds with $p = q+2$.

(2): This follows from (1) (using the observation made in the proof (3)). □

Remark: Let $f \in M(I)$ with $I-Z(f)$ countable. The proof of Theorem 9.5 shows also that each component of $Z(f)$ is eventually periodic, and that each component of $Z_*(f)$ contains exactly one periodic component of $Z(f)$.

Theorem 9.6 Let $f \in M(I)$, C be an f-cycle with period ℓ , and $C_1,\ldots, C_m$ be f-cycles with $\mathrm{int}(C_1),\ldots, \mathrm{int}(C_m)$ disjoint, $\bigcup_{j=1}^{m} C_j \subset C$, and $C - \bigcup_{j=1}^{m} A(C_j,f)$ countable. Then there exists $p \geq 0$ such that:

(i) $\mathrm{per}(C_j) \mid \ell 2^p$ for each $j = 1,\ldots, m$;

(ii) each periodic point in $C - \bigcup_{j=1}^{m} A(C_j,f)$ has a period which divides $\ell 2^p$.

Moreover, each point in $C - \bigcup_{j=1}^{m} A(C_j,f)$ is eventually periodic.

Proof Put $E = C - \bigcup_{j=1}^{m} A(C_j,f)$. Let B be one of the components of C , and let g be the restriction of f^ℓ to B ; so $g \in M(B)$. For $j = 1,\ldots, m$ let $K_j = C_j \cap B$; thus K_j is a g-cycle and $\mathrm{per}(K_j) = \mathrm{per}(C_j)/\ell$. Put $D = B - \bigcup_{j=1}^{m} A(K_j,g)$; then $D \in \mathbf{I}(g)$, and D is countable (since $D \subset E \cup W$, where $W = \{ x \in I : f^k(x) \in \bigcup_{j=1}^{m} \partial C_j$ for some $k \geq 0 \}$). By Proposition 9.1 there exists $q \geq 0$ such that for each $z \in D$ the sequence $\{g^{2^q n}(z)\}_{n\geq 0}$ converges to a periodic point of g in D . Moreover, each point in D is eventually periodic: If $x \in D$ is not eventually periodic and $\lim_{n\to\infty} g^{2^q n}(x) = w$ then, again as in Lemma 9.4, there exists a sink J of g with $w \in \bar{J}$ such that $g^{2qn}(x) \in J$ for all large enough n . Since D is countable it follows that $g^{2qn}(x) \in \mathrm{int}(K_j)$ for some j and some n , and this is not possible because $g^k(x) \in D$ for all $k \geq 0$.

Let $x \in E$ and $v = f^i(x)$, where $0 \leq i < \ell$ is such that $f^i(x) \in B$. If $g^n(v) \in D$ for some $n \geq 0$ then x is eventually periodic and has an eventual period which divides $\ell 2^q$ (since the eventual period of v (with respect to g) divides 2^q). But the only way we can have $g^n(v) \notin D$ for all $n \geq 0$ is if there exists $1 \leq j \leq m$ and $k \geq 0$ such that $g^k(v) = u$, where $u \in K_j \cap \mathrm{Per}(g)$ is an end-point of B with $g^n(u)$ an end-point of K_j for each $n \geq 0$. (Note that this problem can arise because the interior of K_j , which occurs in the definition of $A(K_j,g)$, has to be taken with repect to the relative topology on B .) In this case x is again eventually periodic and has an eventual period which divides $2\ell\,\mathrm{per}(K_j)$. This shows that each point in E is eventually periodic, and each periodic point of f in E has a period which divides ℓN , where N is the least common multiple of $2\,\mathrm{per}(K_1), \ldots, 2\,\mathrm{per}(K_m)$ and 2^q.

Finally, consider the g-cycle K_i, and let L be one of the components of K_i. Let U be the component of $B-D$ containing $\mathrm{int}(L)$; then U is periodic, say with period r, and as in the proof of Theorem 9.5 we have $r \mid 2^{q+1}$. But $U \subset \bigcup_{j=1}^{m} A(K_j,g)$ and the open sets $A(K_1,g), \ldots, A(K_m,g)$ are disjoint; hence, since U is connected, we have $U \subset A(K_i,g)$. Now the same argument shows that $\mathrm{per}(K_i) = r$: Let $L = L_1, \ldots, L_s$ be the components of K_i, and for $k = 1, \ldots, s$ let A_k be the set of points $x \in A(K_i,g)$ such that $g^n(x) \in \mathrm{int}(L_k)$, where $n \geq 0$ is the smallest integer with $g^n(x) \in \bigcup_{j=1}^{s} \mathrm{int}(L_j)$. Then $A_1, \ldots, A_s$ are disjoint open sets with $\bigcup_{k=1}^{s} A_k = A(K_i,g)$, and so $U \subset A_1$. Therefore L is the only component of K_i contained in $\bar{U}$, and this implies that $\mathrm{per}(K_i) = r$.

Theorem 9.6 thus follows with $p = q + 2$. □

Let $f \in M(I)$ and R be an f-register-shift. Recall that R was defined to be tame if there exists an f-cycle K with $R \subset K$ such that $A(K,f) - A(R,f)$ (or, equivalently, $K - A(R,f)$) is countable. The next result shows that the structure of a tame f-register-shift must be very special.

Theorem 9.7 Let $f \in M(I)$ and R be a tame f-register-shift; let K be an f-cycle with $R \subset K$ and $K - A(R,f)$ countable. Then we have $K - A(R,f) = \{ x \in K : x \text{ is eventually periodic} \}$, and if $x \in K \cap \mathrm{Per}(f)$ then x has period $2^\ell\,\mathrm{per}(K)$ for some $\ell \geq 0$. Moreover, if $\{K_n\}_{n\geq1}$ is a generator for R with $K_1 \subset K$ then there exists a strictly increasing sequence $\{q_n\}_{n\geq1}$ of non-negative integers such that $\mathrm{per}(K_n) = 2^{q_n}\,\mathrm{per}(K)$ for each $n \geq 1$.

Proof Let $\{K_n\}_{n\geq1}$ be a generator for R with $K_1 \subset K$ (and note that by Lemma 2.15 such a generator exists); put $K_0 = K$. Then for each $n \geq 0$ we have $A(K_n,f) - A(K_{n+1},f) \subset A(K,f) - A(R,f)$, and so $K_n - A(K_{n+1},f)$ is countable. Thus by Theorem 9.6 $\mathrm{per}(K_{n+1}) = 2^{p_n}\,\mathrm{per}(K_n)$ for some $p_n \geq 0$ (and in fact $p_n > 0$ if $n > 0$), and there exists $r_n \geq 0$ such that each periodic point in $K_n - A(K_{n+1},f)$ has a period which divides $2^{r_n}\,\mathrm{per}(K_n)$. Moreover, each point in $K_n - A(K_{n+1},f)$

is eventually periodic, and hence also each point in $A(K_n,f) - A(K_{n+1},f)$ is eventually periodic. But

$$K - A(R,f) \subset (K_0 - A(K_1,f)) \cup \bigcup_{n\geq 1} (A(K_n,f) - A(K_{n+1},f)) ,$$

and therefore each point in $K - A(R,f)$ is eventually periodic. Thus $K - A(R,f) = \{ x \in K : x$ is eventually periodic $\}$, since it is clear that no point in $A(R,f)$ is eventually periodic. Now we have $\mathrm{per}(K_n) = 2^{q_n} \mathrm{per}(K)$, where $q_n = p_0 + \cdots + p_{n-1}$ for each $n \geq 1$, and the sequence $\{q_n\}_{n\geq 1}$ is strictly increasing. Also

$$K \cap \mathrm{Per}(f) = (K - A(R,f)) \cap \mathrm{Per}(f) \subset \bigcup_{n\geq 0} (K_n - A(K_{n+1},f)) \cap \mathrm{Per}(f) ,$$

because $A(K_n,f) \cap \mathrm{Per}(f) \subset K_n \cap \mathrm{Per}(f)$; hence if $x \in K \cap \mathrm{Per}(f)$ then x has period $2^{\ell}\,\mathrm{per}(K)$ for some $\ell \geq 0$. □

10. EXTENSIONS

Let $f \in \mathbf{M}(I)$ and $\psi \in V(I)$; we say that $[\psi,f]$ is an **extension** of $g \in \mathbf{M}(I)$ if $\psi \circ f = g \circ \psi$; i.e. $[\psi,f]$ is an extension of g if (ψ,g) is a reduction of f . In this section we study the extensions of a mapping $g \in \mathbf{M}(I)$.

For $g \in \mathbf{M}(I)$ we let $\mathbf{E}(g)$ denote the set of countable subsets E of I such that E is g-invariant and $I-E$ is g-almost-invariant. In Theorem 10.5 we show that if $g \in \mathbf{M}(I)$ and $E \subset I$ then there exists an extension $[\psi,f]$ of g with $\psi(I\text{-supp}(\psi)) = E$ if and only if $E \in \mathbf{E}(g)$. (Note that $\psi(I\text{-supp}(\psi))$ is countable for each $\psi \in V(I)$.)

The main application of Theorem 10.5 is to analyse the mappings in $\mathbf{M}_*(I)$; this is really the converse of the analysis of $\mathbf{M}_*(I)$ given in Section 8. Let $f \in \mathbf{M}(I)$ and C be an f-cycle; we call C essentially transitive if there exists a topologically transitive f-cycle $C' \subset C$ such that $C - A(C',f)$ is countable. We call an extension $[\psi,f]$ of $g \in \mathbf{M}(I)$ primary if there exists $z \in \mathrm{Per}(m,g)$ such that $\psi^{-1}(\{z,g(z),\ldots,g^{m-1}(z)\})$ is an essentially transitive f-cycle and

$$\psi(I\text{-supp}(\psi)) \subset \{ y \in I : g^n(y) = z \text{ for some } n \geq 0 \} .$$

Proposition 10.10 shows that if $f \in \mathbf{M}_*(I)$ and (ψ,g) is a primary reduction of f then $[\psi,f]$ is a primary extension of g , and conversely, if $g \in \mathbf{M}_*(I)$ and $[\psi,f]$ is a primary extension of g then $f \in \mathbf{M}_*(I)$ and (ψ,g) is a primary reduction of f . We say that $f \in \mathbf{M}(I)$ is essentially transitive if there exists a topologically transitive f-cycle C such that $I - A(C,f)$ is countable. Theorem 10.12 states that if $f \in \mathbf{M}_*(I)$ is not essentially transitive then there exists an essentially transitive mapping $f_0 \in \mathbf{M}_*(I)$, $n \geq 1$ and $[\psi_1,f_1],\ldots,[\psi_n,f_n]$ such that $[\psi_k,f_k]$ is a primary extension of f_{k-1} for $k = 1,\ldots,n$ and $f = f_n$. We can think of Theorem 10.12 as a "construction kit" for building the mappings in $\mathbf{M}_*(I)$. Note that if $f \in \mathbf{M}(I)$ is essentially transitive then $f \in \mathbf{M}_*(I)$, and by Theorem 8.8 $\mathbf{D}(f) = \{I\}$; it thus follows from Section 6 that f is conjugate to a uniformly piecewise linear mapping.

We now start our study of the extensions of a mapping $g \in \mathbf{M}(I)$ by looking at the properties of $\psi(I\text{-supp}(\psi))$ when $[\psi,f]$ is an extension of g .

Proposition 10.1 Let (ψ,g) be a reduction of $f \in M(I)$. Then $\psi(I\text{-supp}(\psi))$ is g-invariant and $I - \psi(I\text{-supp}(\psi))$ is g-almost-invariant.

Proof Let $x \in \psi(I\text{-supp}(\psi))$; then there exists a component U of $I\text{-supp}(\psi)$ with $\psi(\bar{U}) = \{x\}$. By Theorem 7.4 $I\text{-supp}(\psi)$ is f-almost-invariant, and hence $f(\bar{U}) \subset \bar{V}$ for some component V of $I\text{-supp}(\psi)$. Then $\{g(x)\} = g(\psi(\bar{U})) = \psi(f(\bar{U})) = \psi(\bar{V})$, and therefore $g(x) \in \psi(I\text{-supp}(\psi))$. This shows that $\psi(I\text{-supp}(\psi))$ is g-invariant. Now let $x \in (I-\psi(I\text{-supp}(\psi)))-S(g)$, and choose $z \in I$ with $\psi(z) = x$. Then $z \in \text{supp}(\psi)$ and so by Theorem 7.4 $f(z) \in \text{supp}(\psi)$. We have $\psi(f(z)) = g(x)$, and thus $g(x) \in I-\psi(I\text{-supp}(\psi))$ unless $f(z) \in \bar{U}$ for some component U of $I\text{-supp}(\psi)$. But if $f(z) \in \bar{U}$ then $z \in S(f)\cap\text{supp}(\psi)$ (because $z \notin \bar{V}$ for each component V of $I\text{-supp}(\psi)$), and this is not possible since it implies that $x = \psi(z) \in S(g)$. Hence $I-\psi(I\text{-supp}(\psi))$ is g-almost-invariant. □

Let $g \in M(I)$ and $\psi \in V(I)$; we say that ψ is **g-cocompatible** if $\psi(I\text{-supp}(\psi))$ is g-invariant and $I - \psi(I\text{-supp}(\psi))$ is g-almost-invariant. If $[\psi,f]$ is an extension of $g \in M(I)$ then by Proposition 10.1 ψ is g-cocompatible. We now show that if ψ is g-cocompatible then we can construct $f \in M(I)$ so that $[\psi,f]$ is an extension of g.

Fix $g \in M(I)$ and $\psi \in V(I)$ with ψ g-cocompatible. Put $D = \text{supp}(\psi)$, and let $D_0 = \{ x \in D : x \in \bar{U}$ for some component U of $I-D \}$ and $D_g = \{ x \in D : \psi(x) \in S(g) \}$. D_0 is countable, and D_g is finite (since ψ is at most 2 to 1 on D). Let $x, y \in D_0$ with $x \neq y$; we say that x and y are **neighbours** if they are the end-points of $\bar{U}$ for some component U of $I-D$. If x and y are neighbours then $x \in D_g$ if and only if $y \in D_g$.

Proposition 10.2 There exists a unique continuous mapping $h : D \to D$ with $\psi\circ h = g\circ\psi$. Moreover, we have:

(10.1) If J is a lap of g then h is strictly increasing (resp. strictly decreasing) on $\psi^{-1}(\text{int}(J)) \cap D$ if g is increasing (resp. decreasing) on J.

(10.2) $h(D_0) \subset D_0$.

(10.3) Let $x, y \in D_0$ be neighbours; then

(i) $h(x)$ and $h(y)$ are neighbours if $x \notin D_g$,

(ii) $h(x) = h(y)$ if $x \in D_g$.

Proof We must define h so that $h(x) \in \psi^{-1}(g(\psi(x))) \cap D$ for each $x \in D$. If $x \in D-(D_0 \cup D_g)$ then $\psi(x) \in (I-\psi(I-D))-S(g)$, and hence $g(\psi(x)) \in I - \psi(I-D)$ (since $I - \psi(I-D)$ is g-almost-invariant). Thus $\psi^{-1}(g(\psi(x)))$ ($= \psi^{-1}(g(\psi(x))) \cap D$) consists of a single point, and so in this case we have no choice in the definition of $h(x)$. This shows that there is at most one continuous mapping $h : D \to D$ with $\psi \circ h = g \circ \psi$. ($D - (D_0 \cup D_g)$ is dense in D, since $D_0 \cup D_g$ is countable and D is perfect.) For $x \in D_g$ we define $h(x)$ to be the minimum (resp. maximum) element in $\psi^{-1}(g(\psi(x)))$ when $\psi(x)$ is a local maximum (resp. local minimum) of g. Then $h(x) \in D \cup \{a,b\}$, and in fact it is easy to see that $h(x) \in D$. (If $h(x) = a$ then $g(\psi(x)) = a$, and so $\psi(x)$ is a local minimum of g; hence $\psi^{-1}(g(\psi(x))) = \{a\}$ and $a \in D$.) Finally, consider $x \in D_0-D_g$; then $\psi(x) \notin S(g)$ and thus g is monotone in a neighbourhood of $\psi(x)$. Also there exists a component (y,z) of $I-D$ such that $x \in \{y,z\}$. We define $h(x)$ to be the minimum element in $\psi^{-1}(g(\psi(x)))$ if either $x = y$ and g is increasing in a neighbourhood of $\psi(x)$ or $x = z$ and g is decreasing in a neighbourhood of $\psi(x)$; in the other two cases we define $h(x)$ to be the maximimum element in $\psi^{-1}(g(\psi(x)))$. In all cases we have $h(x) \in D$. This gives us a mapping $h : D \to D$ with $\psi \circ h = g \circ \psi$; we now show that h is continuous. Let $\{x_n\}_{n\geq 1}$ be a sequence from D with $\lim_{n\to\infty} x_n = x$. Then $\lim_{n\to\infty} \psi(h(x_n)) = \lim_{n\to\infty} g(\psi(x_n)) = g(\psi(x))$, and hence $\liminf_{n\to\infty} h(x_n)$ and $\limsup_{n\to\infty} h(x_n)$ both lie in $\psi^{-1}(g(\psi(x)))$. In particular this gives us that $\lim_{n\to\infty} h(x_n) = h(x)$ when $x \in D - (D_0 \cup D_g)$ (since then $\psi^{-1}(g(\psi(x)))$ consists of a single point). Suppose $x \in D_g$, and without loss of generality assume that $\psi(x)$ is a local maximum of g. Then there exists $m \geq 1$ such that either $g(\psi(x_n)) < g(\psi(x))$ or $\psi(x_n) = \psi(x)$ for all $n \geq m$, and therefore $h(x_n) \leq h(x)$ for all $n \geq m$. But $h(x)$ is the minimum element in $\psi^{-1}(g(\psi(x)))$ and $\liminf_{n\to\infty} h(x_n) \in \psi^{-1}(g(\psi(x)))$; thus $\lim_{n\to\infty} h(x_n) = h(x)$. A similar argument deals with the case when $x \in D_0-D_g$. Hence h is continuous. Next let J be a lap of g, and without loss of generality suppose that g is increasing on J. Let $x, y \in \psi^{-1}(\mathrm{int}(J)) \cap D$ with $x < y$. If $(x,y) \not\subset I-D$ then $\psi(x) < \psi(y)$, and from this it follows that $g(\psi(x)) < g(\psi(y))$, and thus also $h(x) < h(y)$. If $(x,y) \subset I-D$ then $x, y \in D_0-D_g$, and so $h(x)$ (resp. $h(y)$) is the minimum (resp. maximum) element in $\psi^{-1}(g(\psi(x)))$. But $\psi(x) \in \psi(I-D)$ and $\psi(I-D)$ is g-invariant; therefore $g(\psi(x)) \in \psi(I-D)$ and hence $\psi^{-1}(g(\psi(x)))$ is a non-trivial interval, i.e. $h(x) < h(y)$. This gives us (10.1). (10.2) and (10.3) follow from the definition of h and the fact that $\psi(I-D)$ is g-invariant. □

Let $g \in M(I)$ and $\psi \in V(I)$ be as before with ψ g-cocompatible, let $h : D \to D$ be as in Proposition 10.2. Consider now a mapping $f : I \to I$ with the following properties:

(10.4) $f(x) = h(x)$ for each $x \in D$.

(10.5) If U is a component of $I-D$ then f is continuous and piecewise monotone on $\bar{U}$ and $f(\bar{U}) \subset \psi^{-1}(g(\psi(\bar{U})))$. (f is piecewise monotone on $[\alpha,\beta]$ means that there exists $N \geq 0$ and $\alpha = d_0 < d_1 < \cdots < d_N < d_{N+1} = \beta$ such that f is strictly monotone on each of the intervals $[d_k,d_{k+1}]$, $k = 0,\ldots, N$.)

(10.6) f is strictly monotone on all but finitely many components of $I-D$.

Note that if U is a component of $I-D$ then, since $\psi(I-D)$ is g-invariant, $\psi^{-1}(g(\psi(\bar{U})))$ is a non-trivial closed interval. Thus by (10.2) and (10.3) there exist mappings f which satisfy (10.4), (10.5) and (10.6).

Proposition 10.3 $f \in M(I)$ and $[\psi,f]$ is an extension of g .

Proof If $x \in D$ then by (10.4) we have $\psi(f(x)) = g(\psi(x))$; if U is a component of $I-D$ and $x \in U$ then by (10.5)

$$\psi(f(x)) \in \psi(f(\bar{U})) \subset \psi(\psi^{-1}(g(\psi(\bar{U})))) = g(\psi(\bar{U})) = \{g(\psi(x))\} ,$$

and so we again have $\psi(f(x)) = g(\psi(x))$. Thus $\psi \circ f = g \circ \psi$. By definition f is continuous on $I-D$, and almost exactly the same proof which showed that h is continuous in Proposition 10.2 shows that f is continuous on D ; therefore f is continuous. Let $U_1,\ldots, U_m$ be the (finitely many) components of $I-D$ on which f is not strictly monotone, and let J be a lap of g . Then by (10.1), and since f is continuous, we have that f is strictly increasing (resp. strictly decreasing) on each component of $\psi^{-1}(\text{int}(J)) - (U_1 \cup \cdots \cup U_m)$ if g is increasing (resp. decreasing) on J . This, together with (10.5), gives us that $f \in M(I)$. □

Let $g \in M(I)$ and $\psi \in V(I)$ be g-cocompatible; let $h : D \to D$ be as in Proposition 10.2 (with $D = \text{supp}(\psi)$). For each component U of $I-D$ let $\sigma_U \in I$

be such that $\psi(U) = \{\sigma_U\}$. Note that if $f : I \to I$ satisfies (10.4), (10.5) and (10.6) and U is a component of $I-D$ with $\sigma_U \in T(g)$ then by (10.3) *(ii)* f cannot be monotone on $\bar{U}$. However, by (10.2) and (10.3) we can find $f : I \to I$ satisfying (10.4), (10.5), (10.7) and (10.8), where:

(10.7) If U is a component of $I-D$ with $\sigma_U \notin T(g)$ then f is strictly monotone on $\bar{U}$.

(10.8) If U is a component of $I-D$ with $\sigma_U \in T(g)$ then f has exactly one turning point in U .

Let f satisfy (10.4), (10.5), (10.7) and (10.8); then by Proposition 10.3 $f \in \mathbf{M}(I)$ and $[\psi,f]$ is an extension of g (since (10.7) and (10.8) imply (10.6)), and it is easy to see that we here have $|T(f)| = |T(g)|$.

Let $g \in \mathbf{M}(I)$ and $\psi \in V(I)$ be g-cocompatible; then by Proposition 10.3 there exists $f \in \mathbf{M}(I)$ such that $[\psi,f]$ is an extension of g . We are thus left with the problem of constructing g-cocompatible elements of $\mathbf{V}(I)$ for a given $g \in \mathbf{M}(I)$, and the next result tells us that this is not difficult to do.

Proposition 10.4 Let $E \subset I$ be countable; then there exists $\psi \in \mathbf{V}(I)$ with $\psi(I\text{-supp}(\psi)) = E$.

Proof This is an easy exercise in real analysis; we give a proof in Section 13 (Proposition 13.6). □

For $g \in \mathbf{M}(I)$ we let $\mathbf{E}(g)$ denote the set of countable subsets E of I such that E is g-invariant and $I-E$ is g-almost-invariant .

Theorem 10.5 Let $g \in \mathbf{M}(I)$ and $E \subset I$; then there exists an extension $[\psi,f]$ of g with $\psi(I\text{-supp}(\psi)) = E$ if and only if $E \in \mathbf{E}(g)$.

Proof This follows from Propositions 10.1, 10.2, 10.3 and 10.4. □

We next analyse the sets in $\mathbf{E}(g)$ for a mapping $g \in \mathbf{M}(I)$. For $z \in I$ let

$$E(z,g) = \{ x \in I : \text{there exist } n , m \geq 0 \text{ such that } g^n(x) = g^m(z) \} ;$$

then $E(z,g) \in \mathbf{E}(g)$, since $E(z,g)$ and $I-E(z,g)$ are both g-invariant. Also if $u, v \in I$ then either $E(u,g) = E(v,g)$ or $E(u,g) \cap E(v,g) = \emptyset$. For $z \in I$ let

$$\mathbf{E}(z,g) = \{ E \in \mathbf{E}(g) : E \neq \emptyset \text{ and } E \subset E(z,g) \} .$$

Lemma 10.6 Let $g \in \mathbf{M}(I)$, $z \in I$ and $E \in \mathbf{E}(z,g)$. Then there exists $W \subset S(g)$ such that $E = \{ x \in E(z,g) : g^n(x) \notin W \text{ for all } n \geq 0 \}$. In particular, $\mathbf{E}(z,g)$ contains at most $2^{|S(g)|}$ elements.

Proof Let $W = S(g) \cap (E(z,g)-E)$; if $x \in E$ then $g^n(x) \notin W$ for all $n \geq 0$, because E is g-invariant. Conversely, consider $x \in E(z,g)-E$; we have $g^n(x) \in E$ for some $n \geq 1$ (since if $y \in E$ then $g^n(x) = g^m(y)$ for some $n, m \geq 0$, and $g^m(y) \in E$), thus let $p = \min\{ n \geq 0 : g^{n+1}(x) \in E \}$. Then $g^p(x) \in W$, because $I-E$ is g-almost-invariant. □

Let $g \in \mathbf{M}(I)$; we put $\mathbf{E}_0(g) = \bigcup_{z \in I} \mathbf{E}(z,g)$. Each member of $\mathbf{E}(g)$ can be written as a countable disjoint union of elements from $\mathbf{E}_0(g)$, since if $E \in \mathbf{E}(g)$ then $E = \bigcup_{z \in E} (E \cap E(z,g))$ and $E \cap E(z,g) \in \mathbf{E}_0(g)$ for each $z \in E$. This also shows that $\mathbf{E}_0(g)$ is the set of indecomposable elements in $\mathbf{E}(g)$: Let $E \in \mathbf{E}(g)$ with $E \neq \emptyset$; then $E \notin \mathbf{E}_0(g)$ if and only if there exist non-empty sets $E_1, E_2 \in \mathbf{E}(g)$ with $E_1 \cap E_2 = \emptyset$ and $E = E_1 \cup E_2$.

We call $E \in \mathbf{E}_0(g)$ **periodic** if $E \cap \mathrm{Per}(g) \neq \emptyset$. If $E \in \mathbf{E}_0(g)$ and $E \subset E(z,g)$ then $E \cap \mathrm{Per}(g) = E(z,g) \cap \mathrm{Per}(g)$; thus E is periodic if and only if $E(z,g)$ is periodic. Moreover, if E is periodic then there exists $m \geq 1$ and $w \in \mathrm{Per}(m,g)$ such that $E \cap \mathrm{Per}(g) = \{w, g(w), \ldots, g^{m-1}(w)\}$. We call $E \in \mathbf{E}_0(g)$ **convergent** if there exists $m \geq 1$ such that $\{g^{mn}(x)\}_{n \geq 0}$ is a convergent sequence for some (and hence for all) $x \in E$. If $E \subset E(z,g)$ then E is convergent if and only if $E(z,g)$ is convergent. Clearly if E is periodic then it is also convergent.

Proposition 10.7 Let $g \in \mathbf{M}(I)$ and $[\psi, f]$ be an extension of g ; let $E = \psi(I-\mathrm{supp}(\psi)) \cap E(z,g)$ for some $z \in \psi(I-\mathrm{supp}(\psi))$ (and so $E \in \mathbf{E}_0(g)$), and put $U = \{ x \in I-\mathrm{supp}(\psi) : \psi(x) \in E \}$.

(1) If E is not convergent then $U - \mathrm{Homt}(f)$ is countable (and so in particular $\mathrm{Homt}(f) \neq \emptyset$).

(2) If E is convergent but not periodic then $U \subset \mathrm{Sink}(f)$ (and so in particular $\mathrm{Sink}(f) \neq \emptyset$).

(3) Suppose E is periodic and let $w \in \mathrm{Per}(m,g)$ be such that $E \cap \mathrm{Per}(g) = \{w, g(w), \ldots, g^{m-1}(w)\}$. Then $C = \psi^{-1}(E \cap \mathrm{Per}(g))$ is an f-cycle with period m, and we have $A(C,f) \subset U$ and $U - A(C,f)$ is countable.

Proof (1): Let V be a component of U and let $\psi(V) = \{x\}$, thus $x \in E$. If $f^j(V) \cap f^k(V) \neq \emptyset$ for some $0 \le j < k$ then $g^j(x) = g^k(x)$, and hence $E \cap \mathrm{Per}(g) \neq \emptyset$. But this is not possible, because E is not periodic, and therefore the intervals $(f^n(V))_{n\ge0}$ are disjoint. Thus for some $p \ge 0$ we have that $f^n(V) \cap S(f) = \emptyset$ for all $n \ge p$, and this means that either $\mathrm{int}(f^p(V))$ is a homterval of f or there exists a sink J of f and $q \ge 0$ such that $f^q(\mathrm{int}(f^p(V))) \subset J$. Suppose the latter holds, and let $m \ge 1$ be such that f^m is increasing on J and $f^m(J) \subset J$. We have $f^{p+q}(V) \subset \bar{J}$, and hence it follows that the sequence $\{g^{mn+p+q}(x)\}_{n\ge0}$ is monotone. This is not possible, because E is not convergent, and so $\mathrm{int}(f^p(V))$ is a homterval of f. Thus $V - \mathrm{Homt}(f)$ is finite, which implies that $U - \mathrm{Homt}(f)$ is countable.

(2): Again let V be a component of U and $\psi(V) = \{x\}$; hence $x \in E$. Since E is convergent there exists $m \ge 1$ and $w \in I$ such that $\lim_{n\to\infty} g^{mn}(x) = w$. Then $g^m(w) = w$, and therefore $g^{mn}(x) \neq w$ for all $n \ge 0$, because E is not periodic. It follows that for some $q \ge 0$ the sequence $\{g^{2mn}(x)\}_{n\ge q}$ converges strictly monotonically to w ; without loss of generality let us assume that this sequence is strictly increasing. Let $u \in V$; then $g^n(x) = \psi(f^n(u))$, and so the sequence $\{f^{2mn}(u)\}_{n\ge q}$ is also strictly increasing; put $v = \lim_{n\to\infty} f^{2mn}(u)$. There thus exists $\varepsilon > 0$ such that f^{2m} is increasing on $[v-\varepsilon, v]$ and $f^{2m}(y) > y$ for all $y \in [v-\varepsilon, v)$, and this implies that $(v-\varepsilon, v)$ is a sink of f. Hence $u \in \mathrm{Sink}(f)$, and therefore $V \subset \mathrm{Sink}(f)$.

(3): $\psi^{-1}(\{g^k(w)\})$ is a non-trivial closed interval for each k, and $\psi^{-1}(\{w\}), \ldots, \psi^{-1}(\{g^{m-1}(w)\})$ are disjoint; thus C is an f-cycle with period m, since $f(\psi^{-1}(A)) \subset \psi^{-1}(g(A))$ for each $A \subset I$. Let V be a component of U and let $\psi(V) = \{x\}$. Then $x \in E$, and so $g^n(x) = w$ for some $n \geq 0$; hence $\psi(f^n(V)) = g^n(\psi(V)) = \{w\}$, and this implies that $f^n(V) \subset C$. Therefore $V - A(C,f)$ is finite, and so $U - A(C,f)$ is countable. Now we clearly have $\mathrm{int}(C) \subset U \subset I\text{-supp}(\psi)$, and thus $A(C,f) \subset I\text{-supp}(\psi)$, because $\mathrm{supp}(\psi)$ is f-invariant. But if $u \in A(C,f)$ then $g^n(\psi(u)) = w$ for some $n \geq 0$, and hence $\psi(u) \in \psi(I\text{-supp}(\psi)) \cap E(w,g) = E$. It follows that $u \in U$; i.e. we have $A(C,f) \subset U$. □

Remark: Let $g \in M(I)$ and $E \in E_0(g)$; the proof of part (2) of Proposition 10.7 shows that if E is convergent but not periodic then $\mathrm{Sink}(g) \neq \emptyset$. Thus if $g \in M_0(I)$ then each element of $E_0(g)$ is either periodic or not convergent.

We can use Proposition 10.7 (1) to construct mappings having homtervals. Let $g \in M(I)$ and $z \in I$; by Theorem 10.5 there exists an extension $[\psi,f]$ of g with $\psi(I\text{-supp}(\psi)) = E(z,g)$, and if $E(z,g)$ is not convergent then by Proposition 10.7 (1) we have $\mathrm{Homt}(f) \neq \emptyset$. It thus only remains to show that $E(z,g)$ is not convergent for suitable $z \in I$, and the next result tells us that this is the case when $I\text{-}Z(g)$ is uncountable.

Proposition 10.8 Let $g \in M(I)$ with $I\text{-}Z(g)$ uncountable. Then $\{ z \in I : E(z,g)$ is not convergent $\}$ is uncountable. Moreover, if $g \in M_0(I)$ then $\{ z \in I : E(z,g)$ is convergent $\}$ is countable.

Proof Suppose first that $g \in M_0(I)$; then, from the remark at the end of the proof of Proposition 10.7, we have

$$\{ z \in I : E(z,g) \text{ is convergent } \} = \{ z \in I : E(z,g) \text{ is periodic } \} .$$

But it is easy to see that $|\mathrm{Fix}(g^n)| \leq |S(g^n)|$ for each $n \geq 1$ (since $\mathrm{Sink}(g) = \emptyset$); hence $\mathrm{Per}(g)$ is countable. Thus $\{ z \in I : E(z,g)$ is periodic $\}$ is countable, because $E(z,g)$ is countable for each $z \in I$. Now suppose that $g \in M(I)$ with $I\text{-}Z(g)$ uncountable; by Proposition 7.9 there then exists a reduction (ψ,h) of g with $h \in M_0(I)$. It is clear that if $E(z,g)$ is convergent then so is $E(\psi(z),h)$, and therefore

$\{ z \in I : E(z,g)$ is not convergent $\}$

$$\supset \psi^{-1}(\{ v \in I : E(v,h) \text{ is not convergent } \}) .$$

But ψ is onto and $\{ v \in I : E(v,h)$ is convergent $\}$ is countable, and so $\{ z \in I : E(z,g)$ is not convergent $\}$ is uncountable. □

Remark: Note that if $g \in \mathbf{M}(I)$ with $I-Z(g)$ countable then by Theorem 9.5 $E(z,g)$ is convergent for all $z \in I$.

Proposition 10.7 (1) can be used to give an example of a mapping $f \in \mathbf{M}(I)$ and a reduction (ψ,g) of f where g has more register-shifts than f . Let $g \in \mathbf{M}_r(I)$, R be the single g-register-shift, and let $z \in R$. By Theorem 10.5 there exists an extension $[\psi,f]$ of g with $\psi(I\text{-supp}(\psi)) = E(z,g)$, and then by Proposition 10.7 (1) we have $(I\text{-supp}(\psi)) - \mathrm{Homt}(f)$ is countable, since $R \cap \mathrm{Per}(g) = \varnothing$ (and $\{ x \in I\text{-supp}(\psi) : \psi(x) \in E(z,g) \} = I\text{-supp}(\psi)$). Now let K be any f-cycle; then $\mathrm{int}(K) \cap \mathrm{supp}(\psi) \neq \varnothing$. (If $\mathrm{int}(C) \cap \mathrm{supp}(\psi) = \varnothing$ then ψ would map each component of K onto a periodic point in $E(z,g)$, and $E(z,g) \cap \mathrm{Per}(g) = \varnothing$.) Therefore by Theorem 7.8 (2) $\psi(K)$ is a g-cycle, and so by Theorem 2.4 $R \subset \psi(K)$. We can thus find $w \in E(z,g) \cap \mathrm{int}(\psi(K))$, and then $J = \psi^{-1}(\{w\})$ is a non-trivial interval with $J \subset K$ and $J - \mathrm{Homt}(f)$ countable. This shows that $K \cap \mathrm{Homt}(f) \neq \varnothing$ for each f-cycle K , and hence there are no f-register-shifts (and also no topologically transitive f-cycles). In particular, g has more register-shifts than f .

The next part of this section is concerned with using extensions to analyse the mappings in $\mathbf{M}_*(I)$; this is really the converse of the analysis of $\mathbf{M}_*(I)$ given in Section 8. The extensions $[\psi,f]$ of a mapping $g \in \mathbf{M}_*(I)$ which now play the main rôle are those for which $\psi(I\text{-supp}(\psi)) = E$ for some periodic $E \in \mathbf{E}_0(g)$.

Proposition 10.9 Let $g \in \mathbf{M}(I)$, $E \in \mathbf{E}_0(g)$ be periodic and let $[\psi,f]$ be an extension of g with $\psi(I\text{-supp}(\psi)) = E$. Put $C = \psi^{-1}(E \cap \mathrm{Per}(g))$, so by Proposition 10.7 (3) C is an f-cycle. Then $\mathrm{supp}(\psi) = \kappa(I-A(C,f))$. If $g \in \mathbf{M}_0(I)$ then $f \in \mathbf{M}_0(I)$ if and only if $C \cap Z(f) = \varnothing$; moreover, if $g \in \mathbf{M}_*(I)$ then $f \in \mathbf{M}_*(I)$ if and only if $C \cap Z(f) = \varnothing$ and C contains no f-register-shifts.

Proof We have $I\text{-supp}(\psi) = \{ x \in I\text{-supp}(\psi) : \psi(x) \in E \}$, and so by Proposition 10.7 (3) $A(C,f) \subset I\text{-supp}(\psi)$ and $(I\text{-supp}(\psi)) - A(C,f)$ is countable; hence $\text{supp}(\psi) = \kappa(I\text{-}A(C,f))$. Now suppose that $g \in \mathbf{M}_0(I)$; if $f \in \mathbf{M}_0(I)$ then of course $C \cap Z(f) = \emptyset$. Conversely, if $C \cap Z(f) = \emptyset$ then also $A(C,f) \cap Z(f) = \emptyset$, and thus $Z(f) = \emptyset$, since by Theorem 7.8 (6) $\kappa(I\text{-}A(C,f)) \cap Z(f) = \emptyset$; i.e. $f \in \mathbf{M}_0(I)$. Finally, suppose $g \in \mathbf{M}_*(I)$, and assume that $C \cap Z(f) = \emptyset$ and C contains no f-register-shifts. Then as before we have $f \in \mathbf{M}_0(I)$, and if R is an f-register-shift then by Proposition 2.10 $R \cap A(C,f) = \emptyset$. But if $R \cap A(C,f) = \emptyset$ then $R \cap \text{supp}(\psi) \neq \emptyset$, and by Theorem 7.8 (4) $\psi(R)$ would then be a g-register-shift. Therefore there are no f-register-shifts, and so $f \in \mathbf{M}_*(I)$. The converse is clear. □

Let $f \in \mathbf{M}(I)$ and C be an f-cycle; we call C **essentially transitive** if there exists a topologically transitive f-cycle $C' \subset C$ such that $C - A(C',f)$ is countable. Note that by Theorem 9.6 there then exist p , $q \geq 0$ such that $\text{per}(C') = 2^p\,\text{per}(C)$ and each periodic point in $C - A(C',f)$ has a period which divides $2^q\,\text{per}(C)$; also each point in $C - A(C',f)$ is eventually periodic.

Let $g \in \mathbf{M}(I)$; we call an extension $[\psi,f]$ of g **primary** if there exists a periodic $E \in \mathbf{E}_0(g)$ such that $\psi(I\text{-supp}(\psi)) = E$ and the f-cycle $C = \psi^{-1}(E \cap \text{Per}(g))$ is essentially transitive.

Proposition 10.10 Let $f \in \mathbf{M}_*(I)$ and (ψ,g) be a primary reduction of f ; then $[\psi,f]$ is a primary extension of g . Conversely, let $g \in \mathbf{M}_*(I)$ and $[\psi,f]$ be primary extension of g ; then $f \in \mathbf{M}_*(I)$ and (ψ,g) is a primary reduction of f .

Proof Let (ψ,g) be a primary reduction of $f \in \mathbf{M}_*(I)$. Then there exists a topologically transitive f-cycle C' such that $\text{supp}(\psi) = \kappa(I\text{-}A(C',f))$. Let B be one of the components of C' ; then $\text{int}(B) \subset I\text{-supp}(\psi)$ and so $\psi(B) = \{z\}$ for some $z \in I$; moreover, $z \in \text{Per}(g)$, since $\psi(f^k(B)) = \{g^k(z)\}$ for each $k \geq 0$. Now put $E = \psi(I\text{-supp}(\psi))$, and let $x \in E$; we thus have $\psi(U) = \{x\}$ for some component U of $I\text{-supp}(\psi)$, and since $U \cap A(C',f) \neq \emptyset$ there exists $n \geq 0$ and $u \in U$ such that $f^n(u) \in B$. Hence $g^n(x) = g^n(\psi(u)) = \psi(f^n(u)) = z$, and so

$E \subset E(z,g)$; i.e. $E \in \mathbf{E}_0(g)$, and E is periodic because $E(z,g)$ is periodic. Let $C = \psi^{-1}(E \cap Per(g)) = \psi^{-1}(\{z,g(z),\dots,g^{m-1}(z)\})$, where m is the period of z ; by Propositions 10.7 (3) and 10.9 C is an f-cycle and $supp(\psi) = \kappa(I-A(C,f))$. But $C' \subset C$, and $C - A(C',f)$ is countable, since $\kappa(I-A(C',f)) = \kappa(I-A(C,f))$; hence C is essentially transitive. This shows that $[\psi,f]$ is a primary extension of g . Conversely, suppose that $g \in \mathbf{M}_*(I)$ and $[\psi,f]$ is a primary extension of g . Then by Proposition 10.9 $f \in \mathbf{M}_*(I)$. Let $E \in \mathbf{E}_0(g)$ be periodic and such that $\psi(I-supp(\psi)) = E$ and the f-cycle $C = \psi^{-1}(E \cap Per(g))$ is essentially transitive. Then there exists a topologically transitive f-cycle $C' \subset C$ with $C - A(C',f)$ countable. Then $A(C,f) - A(C',f)$ is also countable, and therefore $\kappa(I-A(C',f)) = \kappa(I-A(C,f)) = supp(\psi)$. Hence (ψ,g) is a primary reduction of f .

□

Proposition 10.11 (1) Let $g \in \mathbf{M}_*(I)$ and $[\psi,f]$ be a primary extension of g ; suppose there exists an extension $[\psi',f']$ of g and $\theta \in \mathbf{V}(I)$ such that $\psi = \psi'\circ\theta$, and so the following diagram commutes:

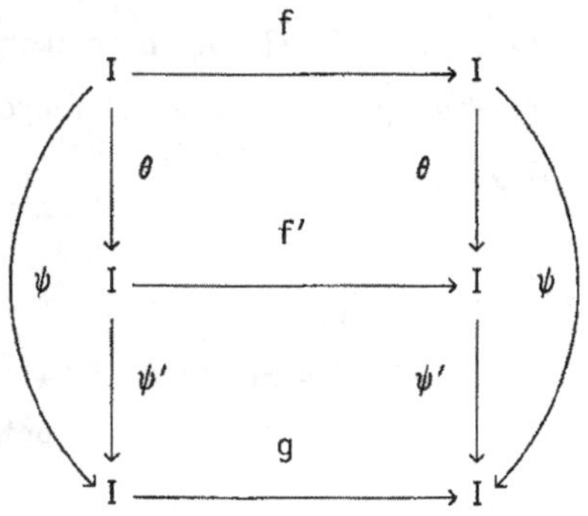

Then either $supp(\psi') = I$, in which case ψ' is a homeomorphism and f' and g are conjugate, or $supp(\psi') = supp(\psi)$, in which case θ is a homeomorphism and f and f' are conjugate.

(2) Let $g \in \mathbf{M}_*(I)$ and $[\psi',f']$ be an extension of g with $f' \in \mathbf{M}_*(I)$ and $supp(\psi') \neq I$. Then there exists a primary extension $[\psi,f]$ of g and $\theta \in \mathbf{V}(I)$ such that $\psi' = \psi\circ\theta$, and hence the following diagram commutes:

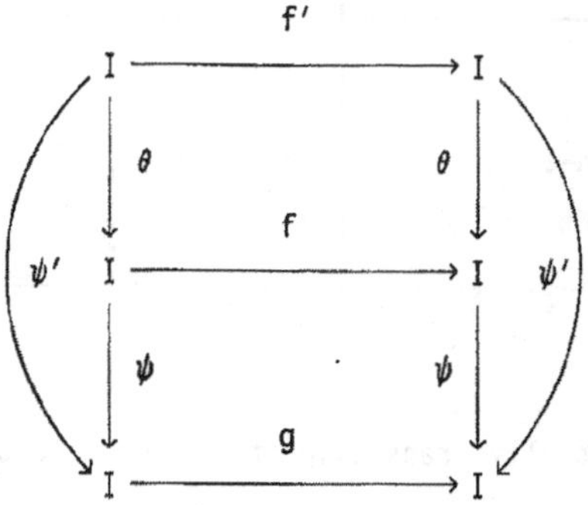

Proof (1): This follows from Proposition 10.10 and Lemma 8.6.

(2): By Theorem 8.8 there exists $n \geq 1$ and $(\psi_1,f_1),\ldots,(\psi_n,f_n)$ such that (ψ_k,f_k) is a primary reduction of f_{k-1} for $k = 1,\ldots,n$ (with $f_0 = f'$) and $(\psi',g) = (\psi_n \circ \cdots \circ \psi_1, f_n)$. Put $\psi = \psi_n$, $f = f_{n-1}$ and $\theta = \psi_{n-1} \circ \cdots \circ \psi_1$ (with $\theta = \mathrm{id}_I$ when $n = 1$). Then $\psi' = \psi \circ \theta$ and (ψ,g) is a primary reduction of f . Thus by Proposition 10.10 $[\psi,f]$ is a primary extension of g . □

We say that $f \in \mathbf{M}(I)$ is **essentially transitive** if there exists a topologically transitive f-cycle C such that $I - A(C,f)$ is countable (i.e. f is essentially transitive when the f-cycle I is essentially transitive). Let $f \in \mathbf{M}(I)$ be essentially transitive; then $f \in \mathbf{M}_*(I)$, and by Theorem 8.8 $\mathbf{D}(f) = \{I\}$; thus if (ψ,g) is a reduction of f then ψ must be a homeomorphism, and so f and g are conjugate. It thus follows from Theorem 6.5 that f is conjugate to a uniformly piecewise linear mapping, (since Lemma 6.4 shows that $h(f) > 0$ when f is essentially transitive.) Let C be the topologically transitive f-cycle; by Theorem 9.6 there then exist p , $q \geq 0$ such that $\mathrm{per}(C) = 2^p$ and each periodic point in $I - A(C,f)$ has a period which divides 2^q ; also each point in $I - A(C,f)$ is eventually periodic.

Theorem 10.12 Let $f \in \mathbf{M}_*(I)$ be not essentially transitive. Then there exists an essentially transitive mapping $f_0 \in \mathbf{M}_*(I)$, $n \geq 1$ and $[\psi_1,f_1],\ldots,[\psi_n,f_n]$ such that $[\psi_k,f_k]$ is a primary extension of f_{k-1} for $k = 1,\ldots,n$ and $f = f_n$.

$$\begin{array}{ccccccccc} I & \xrightarrow{\psi_n} & I & \longrightarrow \cdots \longrightarrow & I & \xrightarrow{\psi_2} & I & \xrightarrow{\psi_1} & I \\ \downarrow{\scriptstyle f = f_n} & & \downarrow{\scriptstyle f_{n-1}} & & \downarrow{\scriptstyle f_2} & & \downarrow{\scriptstyle f_1} & & \downarrow{\scriptstyle f_0} \\ I & \xrightarrow{\psi_n} & I & \longrightarrow \cdots \longrightarrow & I & \xrightarrow{\psi_2} & I & \xrightarrow{\psi_1} & I \end{array}$$

Proof Since f is not essentially transitive there exists a primary reduction (ψ_1', f_1') of f. Suppose for some $m \geq 1$ we have $(\psi_1', f_1'), \ldots, (\psi_m', f_m')$ such that (ψ_k', f_k') is a primary reduction of f_{k-1}' for $k = 1, \ldots, m$ (with $f_0' = f$), and for $k = 1, \ldots, m$ let $D_k = \text{supp}(\psi_k' \circ \cdots \circ \psi_1')$. Then $D_k \in \mathbf{D}(f)$ and $D_m \subset \cdots \subset D_2 \subset D_1$. Moreover, as in the proof of Theorem 8.8 $D_k \neq D_{k+1}$ for each $k = 1, \ldots, m-1$, and therefore $m \leq |\mathbf{D}(f)|$. Now if f_m' is not essentially transitive then there exists a primary reduction (ψ_{m+1}', f_{m+1}') of f_m'. But by Theorem 8.4 $\mathbf{D}(f)$ is finite, and so there exists $n \geq 1$ and $(\psi_1', f_1'), \ldots, (\psi_n', f_n')$ such that (ψ_k', f_k') is a primary reduction of f_{k-1}' for $k = 1, \ldots, n$ (with $f_0' = f$) and f_n' is essentially transitive. For $k = 1, \ldots, n$ let $f_k = f_{n-k}'$ and $\psi_k = \psi_{n-k}'$; then f_0 is essentially transitive, $f_n = f$ and by Proposition 10.10 $[\psi_k, f_k]$ is a primary extension of f_{k-1} for $k = 1, \ldots, n$. □

We end this section by considering a generalization of Theorem 10.12 for mappings in $\mathbf{M}_0(I)$. Let $g \in \mathbf{M}(I)$; we call an extension $[\psi, f]$ of g a **register-shift** extension if there exists a periodic $E \in \mathbf{E}_0(g)$ such that $\psi(I - \text{supp}(\psi)) = E$ and the f-cycle $C = \psi^{-1}(E \cap \text{Per}(g))$ supports an f-register-shift.

Proposition 10.13 Let $f \in \mathbf{M}_0(I)$ and (ψ, g) be a register-shift reduction of f; then $[\psi, f]$ is a register-shift extension of g. Conversely, let $g \in \mathbf{M}_0(I)$ and $[\psi, f]$ be a register-shift extension of g; then $f \in \mathbf{M}_0(I)$ and (ψ, g) is a register-shift reduction of f.

Proof This is almost the same as the proof of Proposition 10.10. □

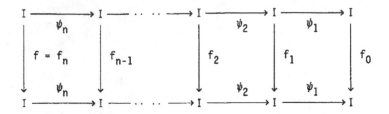

Theorem 10.14 Let $f \in M_0(I)$ and $R_1, \ldots, R_\ell$ be the f-register-shifts; suppose that $I - \bigcup_{i=1}^{\ell} A(R_i, f)$ is uncountable. Then there exists an essentially transitive mapping $f_0 \in M_*(I)$, $n \geq 0$ and $[\psi_1, f_1], \ldots, [\psi_{n+\ell}, f_{n+\ell}]$ such that $[\psi_k, f_k]$ is a primary extension of f_{k-1} for $k = 1, \ldots, n$, $[\psi_j, f_j]$ is a register-shift extension of f_{j-1} for $j = n+1, \ldots, n+\ell$ and $f = f_{n+1}$.

Proof This follows directly from Theorem 10.12, Proposition 8.12 and Proposition 10.13. □

Remark: By Proposition 8.12 $I - \bigcup_{i=1}^{\ell} A(R_i, f)$ will be uncountable if there is a topologically transitive f-cycle or a non-tame f-register-shift.

11. REFINEMENTS

In this section we apply the results of Section 8 to give a fairly complete analysis of the iterates of a mapping $f \in \mathbf{M}_0(I)$. The analysis will be based on the following construct: Let $f \in \mathbf{M}_0(I)$ and $C, C_1, \ldots, C_m$ be f-cycles; we say that $\{C_1,\ldots,C_m\}$ is a **refinement** of C if $\bigcup_{j=1}^{m} C_j \subset C$, $\mathrm{int}(C_1),\ldots, \mathrm{int}(C_m)$ are disjoint and $\mathrm{int}(C - \bigcup_{j=1}^{m} A(C_j,f)) = \varnothing$. Now given an f-cycle C we will try to find a refinement $\{C_1,\ldots,C_m\}$ of C so that the behaviour of f on the f-invariant closed set $C - \bigcup_{j=1}^{m} A(C_j,f)$ is simple enough to be easily described. If this can be done then the problem of studying the behavior of f on C is really reduced to studying the behaviour of f on the f-cycles $C_1,\ldots, C_m$. We then repeat this procedure and look for suitable refinements of the f-cycles $C_1,\ldots, C_m$. Our goal is to start this process with the f-cycle I, and after finitely many steps to end up with f-cycles $K_1,\ldots, K_p$ on which the behaviour of f can be directly described (for example, each K_i should be either topologically transitive or support an f-register-shift).

To realize this programme we must first decide when the behaviour of f on $C - \bigcup_{j=1}^{m} A(C_j,f)$ is to be considered "simple enough". Let $f \in \mathbf{M}_0(I)$ and $\{C_1,\ldots,C_m\}$ be a refinement of an f-cycle C; we say that $\{C_1,\ldots,C_m\}$ is **elementary** (resp. **non-elementary**) if $C - \bigcup_{j=1}^{m} A(C_j,f)$ is countable (resp. uncountable). If $\{C_1,\ldots,C_m\}$ is elementary then Theorem 9.6 describes the behaviour of f on $D = C - \bigcup_{j=1}^{m} A(C_j,f)$: We have that each point in D is eventually periodic, and there exists $p \geq 0$ such that each periodic point of f in D has a period which divides $2^p\,\mathrm{per}(C)$. However, it should be clear that, except in very special cases, we cannot achieve our goal just using elementary refinements, and so we need to consider what is the simplest kind of non-elementary refinement.

Let $f \in \mathbf{M}_0(I)$ and $\{C_1,\ldots,C_m\}$ be a non-elementary refinement of an f-cycle C ; put $D = I - \bigcup_{j=1}^{m} A(C_j,f)$. Then D is closed and f-invariant and I-D is f-almost-invariant; also D is uncountable, and thus by Theorem 7.7 there exists a reduction (ψ,g) of f with $\operatorname{supp}(\psi) = \kappa(D)$. Moreover, $\operatorname{supp}(\psi) \cap \operatorname{int}(C) \neq \emptyset$ and hence by Theorem 7.8 (2) $\psi(C)$ is a g-cycle. Now $\psi(C-D) \subset \psi(I-\operatorname{supp}(\psi))$, and this means that $\psi(C-D)$ is countable; therefore $\psi(C - \bigcup_{j=1}^{m} A(C_j,f)) = \psi(C \cap D) = \psi(C)$ (since $\psi(C \cap D)$ is closed and $\psi(C) = \psi(C \cap D) \cup \psi(C-D)$). Furthermore, it is easy to see that there exists $D_0 \subset D$ with $D - D_0$ countable such that ψ is injective on D_0 . The restriction of f to to $C - \bigcup_{j=1}^{m} A(C_j,f)$ is thus "almost" conjugate to the restriction of g to $\psi(C)$. This suggests that we consider the non-elementary refinement $\{C_1,\ldots,C_m\}$ to be of the simplest kind if the restriction of g to $\psi(C)$ is as simple as possible. But by Theorem 7.8 (6) we have $g \in \mathbf{M}_0(I)$, and also $\psi(C)$ contains no g-register-shifts. (This follows as in the proof of Lemma 8.15, because $\operatorname{int}(C \cap \operatorname{supp}(\psi)) = \emptyset$ implies that $\operatorname{int}(C - \bigcup_{j=1}^{m} A(C_j,f)) = \emptyset$). Hence the simplest possibility is clearly when $\psi(C)$ is a topologically transitive g-cycle, and if this is the case then we say that the refinement $\{C_1,\ldots,C_m\}$ is **transitive**. (Note that this is well-defined: if (ψ',g') is another reduction of f with $\operatorname{supp}(\psi') = \kappa(D)$ then by Theorem 7.4 $\psi'(C)$ is a topologically transitive g′-cycle if and only if $\psi(C)$ is a topologically transitive g-cycle.)

We will be able to carry through the programme outlined above using only elementary and transitive refinements. Let $f \in \mathbf{M}_0(I)$ and $\mathbf{C} = \{C_1,\ldots,C_n\}$, $\mathbf{C'} = \{C'_1,\ldots,C'_m\}$ be two finite sets of f-cycles; we call $\mathbf{C}$ a **refinement** of $\mathbf{C'}$ if there exists $1 \le p \le m$ and a refinement $\mathbf{C''} = \{C''_1,\ldots,C''_q\}$ of the f-cycle C'_p such that $\{C_1,\ldots,C_n\} = \{C'_1,\ldots,C'_{p-1},C''_1,\ldots,C''_q,C'_{p+1},\ldots,C'_m\}$. We also say that $\mathbf{C}$ is **generated** by $(C'_p,\mathbf{C''})$. If $\mathbf{C''}$ is elementary (resp. transitive) then we call $\mathbf{C}$ **elementary** (rep. **transitive**).

Let $\mathbf{C}_0,\ldots,\mathbf{C}_p,\mathbf{C}$ be finite sets of f-cycles; we call $\mathbf{C}_0,\ldots,\mathbf{C}_p$ a **filtration** of C if $\mathbf{C}_0 = \{I\}$, $\mathbf{C}_p = \mathbf{C}$ and $\mathbf{C}_j$ is either an elementary or a transitive refinement of $\mathbf{C}_{j-1}$ for each $j = 1,\ldots p$. If $\mathbf{C} = \{C_1,\ldots,C_m\}$ is a finite set of f-cycles and there exists a filtration of $\mathbf{C}$ then it is easy to see that $\operatorname{int}(C_1),\ldots,\operatorname{int}(C_m)$ are disjoint and $\bigcup_{j=1}^{m} A(C_j,f)$ is dense in I (i.e. $\mathbf{C}$ is a refinement of I). The first result of this section tells us that the converse also holds.

Theorem 11.1 Let $f \in \mathbf{M}_0(I)$ and $C_1,\ldots, C_m$ be f-cycles with $\mathrm{int}(C_1),\ldots, \mathrm{int}(C_m)$ disjoint and $\bigcup_{j=1}^{m} A(C_j,f)$ dense in I. Then there exists a filtration of $C = \{C_1,\ldots,C_m\}$.

Proof This will be a corollary of Theorem 11.5. □

We can immediately apply Theorem 11.1 in conjunction with Theorem 2.4. Let $f \in \mathbf{M}_0(I)$ with topologically transitive f-cycles $C_1,\ldots, C_r$ and f-register-shifts $R_1,\ldots, R_\ell$. For $i = 1,\ldots, \ell$ let K_i be an f-cycle supporting R_i such that $\mathrm{int}(C_1),\ldots, \mathrm{int}(C_r), \mathrm{int}(K_1),\ldots, \mathrm{int}(K_\ell)$ are disjoint. (By Proposition 2.2 and Lemma 2.15 such f-cycles $K_1,\ldots, K_\ell$ exist.) Then by Theorem 2.4 $\bigcup_{j=1}^{r} A(C_j,f) \cup \bigcup_{i=1}^{\ell} A(K_i,f)$ is dense in I, and therefore by Theorem 11.1 there exists a filtration of $\{C_1,\ldots,C_r,K_1,\ldots,K_\ell\}$.

To complete the analysis of the iterates of the mapping f we now describe what happens in a neighbourhood of the f-register-shifts $R_1,\ldots, R_\ell$.

Theorem 11.2 Let $f \in \mathbf{M}_0(I)$ and R be an f-register-shift. Then there exists a decreasing sequence of f-cycles $\{L_n\}_{n\geq 1}$ with L_1 supporting R and $R = \bigcap_{n\geq 1} L_n$ such that $\{L_{n+1}\}$ is either an elementary or a transitive refinement of L_n for each $n \geq 1$.

Remark: If R is tame then $\{L_{n+1}\}$ must be an elementary refinement of L_n for each $n \geq 1$. (See also Theorem 9.7 for more information about the behaviour of f in a neighbourhood of a tame f-register- shift.)

Proof If R is tame then the result is clear. (Just take $\{L_n\}_{n\geq 1}$ to be a generator for R with L_1 supporting R.) If R is not tame then the result will be a corollary of Theorem 11.5. □

Let $f \in \mathbf{M}_0(I)$ with topologically transitive f-cycles $C_1,\ldots, C_r$ and f-register-shifts $R_1,\dots, R_\ell$. For $i = 1,\dots, \ell$ let $\{L_n^i\}_{n\geq 1}$ be a decreasing sequence of f-cycles with L_1^i supporting R_i and $\bigcap_{n\geq 1} L_n^i = R_i$ as in Theorem 11.2. Then the sets $\mathrm{int}(C_1),\dots, \mathrm{int}(C_r), \mathrm{int}(L_1^1),\dots, \mathrm{int}(L_1^\ell)$ must be disjoint. (Let C be an f-cycle with $\mathrm{int}(C) \cap \mathrm{int}(L_1^i) \neq \emptyset$; since we have $\mathrm{int}(L_n^i - A(_{n+1}^i,f)) = \emptyset$ for each $n \geq 1$ it follows that $\mathrm{int}(C) \cap \mathrm{int}(L_n^i) \neq \emptyset$ for all $n \geq 1$. Thus $C \cap R_i \neq \emptyset$ and so by Proposition 2.10 $R_i \subset C$. This shows that $\mathrm{int}(C_j) \cap \mathrm{int}(L_1^i) = \emptyset$ for all $j = 1,\dots, r$ and $i = 1,\dots, \ell$. The same argument gives us that if $\mathrm{int}(L_1^i) \cap \mathrm{int}(L_1^k) \neq \emptyset$ then $\mathrm{int}(L_n^i) \cap \mathrm{int}(L_n^k) \neq \emptyset$ for all $n \geq 1$ and this implies that $R_i \cap R_k \neq \emptyset$; i.e. $i = k$.) Therefore by Theorem 11.1 there exists a filtration of $\{C_1,\dots,C_r,L_1^1,\dots,L_1^\ell\}$.

We will in fact improve somewhat on Theorems 11.1 and 11.2, in that we can restrict the kind of transitive refinement we use. Let $f \in M_o(I)$ and $\{C_1,\dots,C_m\}$ be a non-elementary refinement of an f-cycle C ; let (ψ,g) be a reduction of f with $\mathrm{supp}(\psi) = \kappa(D)$, where $D = I - \bigcup_{j=1}^m A(C_j,f)$. Then, as we have already noted, there exists $D_o \subset D$ with $D - D_o$ countable such that ψ is injective on D_o . But in general ψ need not be very nice on the countable set $D - D_o$, and in particular it need not be finite-to-one there. However, if we choose the f-cycles $C_1,\dots, C_m$ more carefully then it is possible to improve the properties of ψ on $D - D_o$.

Let $f \in M_o(I)$; we call f-cycles $C_1,\dots, C_m$ (with $C_i \neq C_j$ for $i \neq j$) **separated** if whenever $u \in (a,b)$ is an end-point of C_k for some k then each neighbourhood of u contains uncountably many points from $I - \bigcup_{j=1}^m A(C_j,f)$. If $C_1,\dots, C_m$ are separated then in particular they are disjoint. Note that in fact $C_1,\dots, C_m$ are separated if and only if $\bigcup_{j=1}^m \partial C_j \subset \kappa(I - \bigcup_{j=1}^m A(C_j,f))$.

Proposition 11.3 Let $f \in M_o(I)$ and $C_1,\dots, C_n$ be f-cycles; then there exists a unique set of f-cycles $\{K_1,\dots,K_m\}$ such that $K_1,\dots, K_m$ are separated, $\bigcup_{i=1}^n C_i \subset \bigcup_{j=1}^m K_j$ and $\bigcup_{j=1}^m K_j - \bigcup_{i=1}^n A(C_i,f)$ is countable. We call $\{K_1,\dots,K_m\}$ the **separated hull** of $\{C_1,\dots,C_n\}$.

Proof Put $D = \kappa(I - \bigcup_{i=1}^{n} A(C_i,f))$, and let $U = I-D$; thus U is f-almost-invariant. Consider a periodic component J of U with period q , and for $k = 0,\dots, q-1$ let J_k be the component of U with $f_k(\bar{J}) \subset \overline{J_k}$. Then the intervals $\overline{J_0},\dots, \overline{J_{q-1}}$ are disjoint (since D is perfect), and hence $K = \bigcup_{k=0}^{q-1} \overline{J_k}$ is an f-cycle with period q ; also $\operatorname{int}(K) \cap T(f) \neq \emptyset$, because $f \in \mathbf{M}_0(I)$. Let $K_1,\dots, K_m$ be the (different) f-cycles obtained in this way; it is clear that $K_1,\dots, K_m$ are disjoint and $m \le |T(f)|$. Also, since each component of U is eventually periodic, we have $\kappa(I - \bigcup_{j=1}^{m} A(K_j,f)) = D$, and therefore $\bigcup_{j=1}^{m} \partial K_j \subset D = \kappa(I - \bigcup_{j=1}^{m} A(K_j,f))$; i.e. $K_1,\dots, K_m$ are separated. Now $\bigcup_{i=1}^{n} \operatorname{int}(C_i) \subset U$, and from this it immediately follows that $\bigcup_{i=1}^{n} C_i \subset \bigcup_{j=1}^{m} K_j$. Also $\bigcup_{j=1}^{m} K_j - \bigcup_{i=1}^{n} A(C_i,f)$ is countable because $D = \kappa(I - \bigcup_{i=1}^{n} A(C_i,f))$. Suppose $\{L_1,\dots,L_p\}$ is another set of f-cycles with these properties; then we must again have $\kappa(I - \bigcup_{j=1}^{p} A(L_j,f)) = D$, and from this it is easy to see that $\{L_1,\dots,L_p\} = \{K_1,\dots,K_m\}$. □

Let $f \in \mathbf{M}_0(I)$ and $C_1,\dots, C_m$ be separated f-cycles. Note that if $\{C_1,\dots,Cm\} \neq \{I\}$ then $I - \bigcup_{j=1}^{m} A(C_j,f)$ is uncountable.

Proposition 11.4 Let $f \in \mathbf{M}_0(I)$ and $C_1,\dots, C_m$ be separated f-cycles with $\{C_1,\dots,C_m\} \neq \{I\}$; let (ψ,g) be a reduction of f with $\operatorname{supp}(\psi) = \kappa(I - \bigcup_{j=1}^{m} A(C_j,f))$. Then ψ is at most N to 1 on $I - \bigcup_{j=1}^{m} A(C_j,f)$, where $N = 2^{(|T(f)|-m)} + 1$.

Proof Since $C_1,\dots, C_m$ are separated it is easily checked that $\bigcup_{j=1}^{m} \operatorname{int}(C_j)$ is the union of the periodic components of $I-\operatorname{supp}(\psi)$. Let $x < y$ with $\psi(x) = \psi(y)$; then $(x,y) \subset U$ for some component U of $I-\operatorname{supp}(\psi)$. Now if $z \in U$ then $f^n(z) \in \bigcup_{j=1}^{m} C_j$ for some $n \ge 0$; let $p \ge 0$ be the smallest such integer. It follows that if $z \in I - \bigcup_{j=1}^{m} A(C_j,f)$ then $f^k(z) = w \in T(f)$ for some $0 \le k < p$ (and $f^p(z) \in \bigcup_{j=1}^{m} \partial C_j$). Therefore each component of $I\text{-supp}(\psi)$ can contain at most $2^{(|T(f)|-m)} - 1$ points from $I - \bigcup_{j=1}^{m} A(C_j,f)$ (since $w \notin \bigcup_{j=1}^{m} \text{int}(C_j)$ and there are at least m points from $T(f)$ in $\bigcup_{j=1}^{m} \text{int}(C_j)$). Thus ψ is at most N to 1 on $I - \bigcup_{j=1}^{m} A(C_j,f)$. □

Remark: Let $f \in \mathbf{M}_0(I)$ and $C_1,\ldots, C_m$ be separated f-cycles; suppose that $m = |T(f)|$. Then the proof of Proposition 11.4 shows that $I - \bigcup_{j=1}^{m} A(C_j,f)$ is already perfect. Thus if $\{C_1,\ldots,C_m\} \neq I$ then there exists a reduction (ψ,g) of f with $\text{supp}(\psi) = I - \bigcup_{j=1}^{m} A(C_j,f)$, and ψ is at most 2 to 1 on $I - \bigcup_{j=1}^{m} A(C_j,f)$. Note that this situation always occurs whenever $|T(f)| = 1$.

Let $f \in \mathbf{M}_0(I)$ and $\mathbf{C} = \{C_1,\ldots,C_n\}$ be a refinement of an f-cycle C ; we say that $\mathbf{C}$ is **separated** if the f-cycles $C_1,\ldots, C_n$ are separated. Similarly, if $\mathbf{C}$ and $\mathbf{C'}$ are two finite sets of f-cycles and $\mathbf{C}$ is a refinement of $\mathbf{C'}$ generated by $(C,\mathbf{C''})$, then we say that $\mathbf{C}$ is **separated** if $\mathbf{C''}$ is a separated refinement of C .

Theorem 11.5 Let $f \in \mathbf{M}_0(I)$ and $\mathbf{C} = \{C_1,\ldots,C_n\}$, $\mathbf{K} = \{K_1,\ldots K_m\}$ be two sets of separated f-cycles with $\bigcup_{i=1}^{m} K_i \subset \bigcup_{j=1}^{n} C_j$ and $\text{int}(\bigcup_{j=1}^{n} C_j - \bigcup_{i=1}^{m} A(K_i,f)) = \emptyset$. Then there exist finite sets of f-cycles $\mathbf{C}_0,\ldots, \mathbf{C}_p$ with $\mathbf{C}_0 = \mathbf{C}$ and $\mathbf{C}_p = \mathbf{K}$ such that for each $j = 1,\ldots, p$ either $\mathbf{C}_{j-1}$ is the separated hull of $\mathbf{C}_j$ or $\mathbf{C}_j$ is a separated transitive refinement of $\mathbf{C}_{j-1}$.

Proof Later. □

We now show how Theorem 11.1 and Theorem 11.2 follow from Theorem 11.5.

Lemma 11.6 Let $f \in M_0(I)$ and $C_1,\dots, C_n$ be f-cycles with $\text{int}(C_1),\dots, \text{int}(C_n)$ disjoint; let $\{K_1,\dots,K_m\}$ be the separated hull of $\{C_1,\dots,C_n\}$. Then there exist finite sets of f-cycles $\mathbf{C}_1,\dots, \mathbf{C}_p$ with $\mathbf{C}_0 = \{K_1,\dots,K_m\}$ and $\mathbf{C}_p = \{C_1,\dots,C_n\}$ such that $\mathbf{C}_j$ is an elementary refinement of $\mathbf{C}_{j-1}$ for each $j = 1,\dots, p$.

Proof This is clear. □

Theorem 11.1 follows immediately from Theorem 11.5 and Lemma 11.6; moreover, if $C_1,\dots, C_m$ are as in Theorem 11.1 then there exists a filtration of $\mathbf{C} = \{C_1,\dots,C_m\}$ such that each transitive refinement which occurs is separated.

We next consider Theorem 11.2. Let $f \in M_0(I)$ and R be a non-tame f-register-shift. Let $\{K_n\}_{n\geq 1}$ be a generator for R such that K_1 supports R. By replacing $\{K_n\}_{n\geq 1}$ if necessary with a sub-sequence we can assume that $K_n - A(K_{n+1},f)$ is uncountable for each $n \geq 1$. Let $\{L_n\}$ be the separated hull of $\{K_n\}$. (It is clear that the separated hull of a single f-cycle consists of a single f-cycle.) Then $L_{n+1} \subset L_n$ and $\text{int}(L_n - A(L_{n+1},f)) = \varnothing$ for each $n \geq 1$ (since $\text{int}(A(K_n,f) - A(K_{n+1},f)) = \varnothing$ and $L_n - A(L_{n+1},f) \subset (L_n-A(K_n,f)) \cup (A(K_n,f)-A(K_{n+1},f))$). Also $L_{n+1} \neq L_n$, because $K_n - A(K_n,f)$ is uncountable. Let $n \geq 1$; we can apply Theorem 11.5 to the separated sets of f-cycles $\{L_n\}$ and $\{L_{n+1}\}$, and it is easy to see that the finite sets of f-cycles which we then obtain each consists of a single f-cycle. There thus exist f-cycles $L_n = C_0^n \supset C_1^n \supset \cdots \supset C_{p_n}^n = L_{n+1}$ such that for each $j = 1,\dots, p_n$ either $\{C_{j-1}^n\}$ is the separated hull of $\{C_j^n\}$ (in which case $\{C_j^n\}$ is an elementary refinement of C_{j-1}^n) or $\{C_j^n\}$ is a separated transitive refinement of C_{j-1}^n. Therefore Theorem 11.2 will follow if we have $\bigcap_{n\geq 1} L_n = R$. Now clearly $\bigcap_{n\geq 1} L_n \supset R$, and so by Propositions 2.2 (7) and 2.11 it is enough to show that $\lim_{n\to\infty} \text{per}(L_n) = \infty$. But if $\lim_{n\to\infty} \text{per}(L_n) = N < \infty$ then $L = \bigcap_{n\geq 1} L_n$ is an f-cycle with period N, and

$$L - A(R,f) \;=\; \bigcap_{n\geq 1} L_n - \bigcap_{n\geq 1} A(K_n,f) \;\subset\; \bigcup_{n\geq 1} (L_n-A(K_n,f)) ,$$

which is countable; this is not possible because R is not tame. Hence we have $\bigcap_{n\geq 1} L_n = R$.

Proof of Theorem 11.5 Let $\mathbf{C} = \{C_1,\dots,C_n\}$ and $\mathbf{K} = \{K_1,\dots,K_m\}$ be as in the statement of the theorem; we can assume that $\mathbf{C} \neq \mathbf{K}$.

Lemma 11.7 There exists a finite set of f-cycles $\mathbf{L} = \{L_1,\dots,L_\ell\}$ with $\bigcup_{k=1}^{\ell} L_k \subset \bigcup_{j=1}^{n} C_j$ such that $\mathbf{K}$ is a separated transitive refinement of $\mathbf{L}$.

Proof Since $\mathbf{C} \neq \mathbf{K}$ we have that $\bigcup_{j=1}^{n} C_j - \bigcup_{i=1}^{m} A(K_i,f)$ is uncountable, and so in particular there exists a reduction (ψ,g) of f with $\mathrm{supp}(\psi) = \kappa(I - \bigcup_{i=1}^{m} A(K_i,f))$. Then $\mathrm{supp}(\psi) \cap \bigcup_{j=1}^{n} \mathrm{int}(C_j) \neq \varnothing$ and thus by Theorem 7.8 (2) $\psi(C_p)$ is a g-cycle for some p . Now $\mathrm{int}(\mathrm{supp}(\psi) \cap C_p) = \varnothing$, and hence as in the proof of Lemma 8.15 $\psi(C_p)$ contains no g-register-shifts; also by Theorem 7.8 (6) $g \in \mathbf{M}_0(I)$, and therefore Theorem 2.4 implies there exists a topologically transitive g-cycle $C' \subset \psi(C_p)$. Put $C = \psi^{-1}(C')$; then $C \subset C_p$, because $\partial C_p \subset \mathrm{supp}(\psi)$. (This follows from the fact that $C_1,\dots, C_n$ are separated and $\kappa(I - \bigcup_{j=1}^{n} A(C_j,f)) \subset \mathrm{supp}(\psi)$.) For each $i = 1,\dots, m$ either $\psi(K_i) \subset C'$ or $\psi(K_i) \subset I\text{-}C'$ (since $\psi(K_i)$ is a finite periodic orbit), and without loss of generality we can asssume that $\psi(K_i) \subset C'$ for $i = 1,\dots, q$ and $\psi(K_i) \subset I\text{-}C'$ for $i = q+1,\dots, m$. Then $K_i \subset C$ for $i = 1,\dots, q$ and $K_i \subset I\text{-}C$ for $i = q+1,\dots, m$. Also $\mathrm{int}(C - \bigcup_{i=1}^{q} A(K_i,f)) = \varnothing$, because $\mathrm{int}(\bigcup_{j=1}^{n} C_j - \bigcup_{i=1}^{m} A(K_i,f)) = \varnothing$. Thus $q \geq 1$ and $\{K_1,\dots,K_q\}$ is a refinement of C ; moreover, the f-cycles $K_1,\dots, K_q$ are clearly separated, and hence the refinement is separated. Let (ψ',g') be a reduction of f with $\mathrm{supp}(\psi') = \kappa(I - \bigcup_{i=1}^{q} A(K_i,f))$. Then $\mathrm{supp}(\psi) \subset \mathrm{supp}(\psi')$ and so by Theorem 7.4 there exists $\theta \in V(I)$ with $\psi = \theta \circ \psi'$. Now θ must map $\psi'(C)$ homeomorphically onto $\psi(C) = \psi(\psi^{-1}(C')) = C'$. (Suppose we had x , $y \in \psi'(C)$ with $x < y$ and $\theta(x) = \theta(y)$; choose u , $v \in C$ with $\psi'(u) = x$ and $\psi'(v) = y$. Then $\psi(u) = \psi(v)$, and this implies that $(u,v) \cap (I - \bigcup_{i=1}^{q} A(K_i,f))$ is countable, since $(u,v) \subset C$. Therefore $\psi'(u) = \psi'(v)$, which is not possible, because $x \neq y$.) But C' is topologically transitive, and thus $\psi'(C)$ is a topologically transitive g-cycle. This means that $\{K_1,\dots,K_q\}$ is a separated transitive refinement of C ;

hence $\mathbf{K}$ is a separated transitive refinement of $\mathbf{L} = \{C, K_{q+1}, \ldots, K_m\}$ and $C \cup \bigcup_{i=q+1}^{m} K_i \subset \bigcup_{j=1}^{n} C_j$. □

Let $\mathbf{L}$ be as in Lemma 11.7 and let $\mathbf{G} = \{G_1, \ldots, G_p\}$ be the separated hull of $\mathbf{L}$. Then $\bigcup_{i=1}^{m} K_i \subset \bigcup_{q=1}^{p} G_q \subset \bigcup_{j=1}^{n} C_j$, and $\mathrm{int}(\bigcup_{j=1}^{n} C_j - \bigcup_{q=1}^{p} A(G_q, f)) = \varnothing$. Thus if $\mathbf{G} \neq \mathbf{C}$ then we can apply Lemma 11.7 to $\mathbf{C}$ and $\mathbf{G}$. Repeating this procedure generates separated sets of f-cycles $\mathbf{G}_u = \{G_1^u, \ldots, G_{p_u}^u\}$, $u = 1, \ldots, N$, such that

$$\bigcup_{i=1}^{m} K_i \subset \bigcup_{q=1}^{p_1} G_q^1 \subset \cdots \subset \bigcup_{q=1}^{p_N} G_q^N \subset \bigcup_{j=1}^{n} C_j ,$$

and Theorem 11.5 will follow if we have $\mathbf{G}_N = \mathbf{C}$ for some $N \geq 1$. Put $D = \kappa(I - \bigcup_{j=1}^{n} A(C_j, f))$, $E = \kappa(I - \bigcup_{i=1}^{m} A(K_i, f))$ and $D_u = \kappa(I - \bigcup_{q=1}^{p_u} A(G_q^u, f))$; thus each of these sets is either empty or an element of $\mathbf{D}(f)$. We have $E \supset D_1 \supset \cdots \supset D_N \supset D$, and if $i \neq j$ then $D_i \neq D_j$ (since $\mathbf{G}_i \neq \mathbf{G}_j$). Hence in particular $E \in \mathbf{D}(f)$. But $\mathrm{int}(\bigcup_{j=1}^{n} A(C_j, f) - \bigcup_{i=1}^{m} A(K_i, f)) = \varnothing$, and this implies that $\mathrm{int}(E-D) = \varnothing$, and thus also $\mathrm{int}(E-D_u) = \varnothing$ for each u . Therefore by Proposition 8.5 we must have $D_N = D$ for some $N \geq 1$, and then $\mathbf{G}_N = \mathbf{C}$.

This completes the proof of Theorem 11.5. □

12. MAPPINGS WITH ONE TURNING POINT

In this section we analyse the structure of a mapping $f \in M_0(I)$ with one turning point. Without loss of generality we can assume that the mapping f takes on a maximum at its turning point. Moreover, it will be convenient to assume also that $f(\{a,b\}) \subset \{a,b\}$, i.e. that $f(a) = f(b) = a$. Note that it is always possible to reduce things to this case by extending the domain of definition of f to a larger interval $[a',b']$.

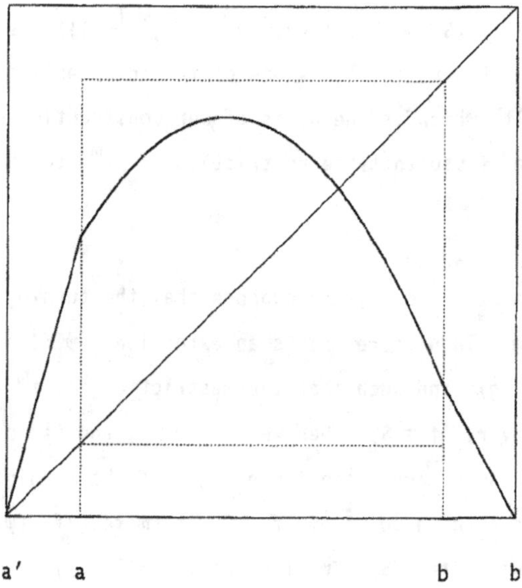

Finally, again without loss of generality, we will assume that $[a,b] = [0,1]$; thus in this section I will always denote the interval $[0,1]$. We let **S** denote the set of mappings $f \in M_0(I)$ having exactly one turning point and for which $f(0) = f(1) = 0$. Let $f \in$ **S** ; then by Theorem 2.4 we have that exactly one of the following holds:

(12.1) There exists a single topologically transitive f-cycle C ; in this case $A(C,f)$ is dense in I.

(12.2) There exists a single f-register-shift R, and in this case $A(R,f)$ is dense in I.

Let S_t (resp. S_r) denote the set of mappings $f \in S$ for which (12.1) (resp. (12.2)) holds.

We start by considering the mappings in S_t , and introduce a procedure for constructing each element of S_t out of finitely many uniformly piecewise linear mappings. Let $g \in S_t$ and suppose the turning point γ of g is periodic with period m . Then by Theorem 10.5 there exists an extension $[\psi,f]$ of g with $\psi(\text{I-supp}(\psi)) = E(\gamma,g)$, and since γ is periodic we have

$$E(\gamma,g) = \{ x \in I : g^n(x) = \gamma \text{ for some } n \geq 0 \} .$$

Moreover, by Proposition 10.7 (3) $C = \psi^{-1}(\{\gamma,g(\gamma),\ldots,g^{m-1}(\gamma)\})$ is an f-cycle with period m ; in particular, $B = \psi^{-1}(\{\gamma\})$ is one of the components of C , and hence $f^m(B) \subset B$. The next result, which is the basis of our construction procedure, shows that we can in fact choose f so that the restriction of f^m to B is conjugate to a given element $q \in S$.

Proposition 12.1 Let $g \in S_t$, $q \in S$, and suppose that the turning point γ of g is periodic with period m . Then there exists an extension $[\psi,f]$ of g with $f \in S$, $\psi(\text{I-supp}(\psi)) = E(\gamma,g)$ and such that the restriction of f^m to $\psi^{-1}(\{\gamma\})$ is conjugate to q . Moreover, $f \in S_t$ whenever $q \in S_t$, and if q is essentially transitive then $[\psi,f]$ is a primary extension of g ; if the turning point of q is periodic then so is the turning point of f . Furthermore, if $[\psi_1,f_1]$ is another extension of g with $f_1 \in S$, $\psi_1(\text{I-supp}(\psi_1)) = E(\gamma,g)$ and such that the restriction of f_1^m to $\psi_1^{-1}(\{\gamma\})$ is conjugate to q , then there exists a homeomorphism $\theta \in V(I)$ with $\psi_1 = \psi\circ\theta$ and $f\circ\theta = \theta\circ f_1$.

Proof Later. □

Let g , q and $[\psi,f]$ be as in Proposition 12.1; we then call $[\psi,f]$ a q **extension** of g . If $[\psi_1,f_1]$ is another q extension of g then by Proposition 12.1 there exists a homeomorphism $\theta \in V(I)$ with $\psi_1 = \psi\circ\theta$ and $f\circ\theta = \theta\circ f_1$.

Proposition 12.2 Let g, $g_1 \in \mathbf{S}_t$, q, $q_1 \in \mathbf{S}$ with g and g_1 conjugate and q and q_1 conjugate. Suppose that the turning point of g (and thus also of g_1) is periodic, and let $[\psi,f]$ be a q extension of g and $[\psi_1,f_1]$ be a q_1 extension of g_1 . Then f and f_1 are conjugate.

Proof Since g and g_1 are conjugate there exists a homeomorphism $\sigma \in V(I)$ with $g \circ \sigma = \sigma \circ g_1$. It is then easy to check that $[\sigma \circ \psi_1, f_1]$ is a q extension of g , and so by Proposition 12.1 there exists a homeomorphism $\theta \in V(I)$ with $\sigma \circ \psi_1 = \psi \circ \theta$ and $f \circ \theta = \theta \circ f_1$. In particular, f and f_1 are conjugate. □

Now let $n \geq 0$ and $h_0, h_1, \ldots, h_n \in \mathbf{S}_t$ be essentially transitive, and suppose that for each $k \neq n$ the turning point of h_k is periodic. Then by Proposition 12.1 we can define $f_0, \ldots, f_n \in \mathbf{S}_t$ by letting f_0 be conjugate to h_0 and for $k = 1, \ldots, n$ letting f_k be an h_k extension of f_{k-1} . Let $f \in \mathbf{S}_t$ be conjugate to f_n ; we say then that $h_0, \ldots, h_n$ is an **extension sequence** for f . If $h_0, \ldots, h_n$ (resp. $h'_0, \ldots, h'_n$) is an extension sequence for $f \in \mathbf{S}_t$ (resp. $f' \in \mathbf{S}_t$) and h_k and h'_k are conjugate for each $k = 0, \ldots, n$ then by Proposition 12.2 f and f' are conjugate. In Theorems 12.3 and 12.4 we show that each element of $\mathbf{S}_t$ possesses an (up to conjugacy) unique extension sequence; i.e. if $f \in \mathbf{S}_t$ then there exists an extension sequence $h_0, \ldots, h_n$ for f , and if $h'_0, \ldots, h'_m$ is another extension sequence for f then $m = n$ and h_k and h'_k are conjugate for each $k = 0, \ldots, n$.

For $\beta \in (1,2]$ let u_β be the unique element of $\mathbf{S}$ which is uniformly piecewise linear with slope β ; (see the picture on the following page). Thus in fact we have

$$u_\beta(x) = \begin{cases} \beta x & \text{if } 0 \leq x \leq \frac{1}{2} , \\ \beta - \beta x & \text{if } \frac{1}{2} \leq x \leq 1 . \end{cases}$$

By Theorem 6.14 we have $h(u_\beta) = \log \beta$ for each $\beta \in (1,2]$. In particular, if α, $\beta \in (1,2]$ with $\alpha \neq \beta$ then u_α and u_β are not conjugate. (If f, $g \in \mathbf{M}(I)$ are conjugate then $\ell(f^n) = \ell(g^n)$ for all $n \geq 1$, and so $h(f) = h(g)$.)

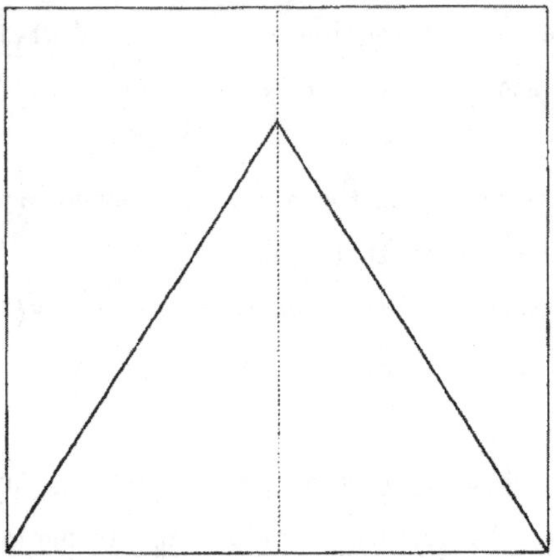

Let $f \in \mathbf{S}_t$ be essentially transitive; then by Theorem 8.8 $\mathbf{D}(f) = \{I\}$, and so by Lemma 6.4 and Theorem 6.5 there exists a homeomorphism $\theta \in \mathbf{V}(I)$ and a uniformly piecewise linear mapping $g \in \mathbf{M}(I)$ with slope $\beta = \exp(h(f)) > 1$ such that $\theta \circ f = g \circ \theta$. It follows that $g = u_\beta$, since $g \in \mathbf{S}$; also it is clear that $\beta \le 2$. Thus if $f \in \mathbf{S}_t$ is essentially transitive then there exists a unique $\beta \in (1,2]$ such that f and u_β are conjugate. In Proposition 12.5 we show that, conversely, the mappings u_β, $1 < \beta \le 2$, are all essentially transitive.

From Theorems 12.3, 12.4 and 6.5 we have that for each $f \in \mathbf{S}_t$ there exists a unique finite sequence $\beta_0, \ldots, \beta_n$, with $n \ge 0$ and $\beta_k \in (1,2]$ for each $k = 0, \ldots, n$, such that $u_{\beta_0}, \ldots, u_{\beta_n}$ is an extension sequence for f. We call $\beta_0, \ldots, \beta_n$ the **characteristic sequence** of f. It then follows from Proposition 12.2 that two elements of $\mathbf{S}_t$ are conjugate if and only if they have the same characteristic sequence.

Let $\mathbf{P} = \{\, \beta \in (1,2]$: the turning point of u_β is periodic $\}$. If $\beta_0, \ldots, \beta_n$ is the characteristic sequence of some element of $\mathbf{S}_t$ then we must have $\beta_k \in \mathbf{P}$ for each $k \ne n$. On the other hand, if $\beta_0, \ldots, \beta_n$ is any sequence from $(1,2]$ with $\beta_k \in \mathbf{P}$ for each $k \ne n$ then by Propositions 12.1 and 12.5 $\beta_0, \ldots, \beta_n$ is the characteristic sequence of some element of S_t . In Proposition 12.6 we show that **P** is countable, and that in fact each $\beta \in \mathbf{P}$ is an algebraic number (i.e. β is a root of some polynomial with integer coefficients).

Proof of Proposition 12.1 Let $\psi \in V(I)$ with $\psi(I\text{-supp}(\psi)) = E(\gamma,g)$ and put $D = \text{supp}(\psi)$; by Proposition 10.2 there then exists a unique continuous mapping $h : D \to D$ with $\psi \circ h = g \circ \psi$, and (10.1), (10.2) and (10.3) hold. Let U be a component of $I\text{-}D$; then, since $g(\{0,1\}) \subset \{0,1\}$, we have $U \subset (0,1)$. (If $\psi(U) = \{z\}$ then $z \in E(\gamma,g)$ and so $g^n(z) = \gamma$ for some $n \geq 0$; thus $z \notin \{0,1\}$.) Hence, putting $U = (x,y)$, we have x , $y \in D_0$. Now let $\psi^{-1}(\{\gamma\}) = [c,d]$, in particular (c,d) is a component of $I\text{-}D$; we first define f on the components of $I\text{-}D$ not equal to (c,d) : If $U = (x,y) \neq (c,d)$ is such a component then by (10.3) $h(x) \neq h(y)$ and so we can define f to be continuous and strictly monotone on $[x,y]$ with $f(x) = h(x)$ and $f(y) = h(y)$. This leaves us to define f on (c,d) : By (10.3) $h(c) = h(d)$, and therefore we also have $h^m(c) = h^m(d) \in \{c,d\}$; there thus exists $q' \in \mathbf{M}([c,d])$ conjugate to q such that $q'(c) = h^m(c) = h^m(d) = q'(d)$. But if $J = \psi^{-1}(\{g(\gamma)\})$ then f^{m-1} maps J homeomorphically onto $[c,d]$, and hence we can define $f : [c,d] \to J$ by $f = (f^{m-1})^{-1} \circ q'$. Then by Proposition 10.3 $f \in \mathbf{M}(I)$ and $[\psi,f]$ is an extension of g ; also the restriction of f^m to $[c,d]$ is equal to q' , which by definition is conjugate to q . Moreover, by Proposition 10.9 we have $f \in \mathbf{M}_0(I)$, and so $f \in \mathbf{S}$; if the turning point of q is periodic with period ℓ then it is clear that the turning point of f is periodic with period ℓm . Furthermore, if $q \in \mathbf{S}_t$ then again using Proposition 10.9 we have $f \in \mathbf{M}_*(I)$, and so $f \in \mathbf{S}_t$; if q is essentially transitive then by construction $[\psi,f]$ is a primary extension of g .

Now let $[\psi_1,f_1]$ be another extension of g with $f_1 \in \mathbf{S}$, $\psi_1(I\text{-supp}(\psi_1)) = E(\gamma,g)$ and such that the restriction of f_1^m to $\psi_1^{-1}(\{\gamma\})$ is conjugate to q . Put $E = E(\gamma,g)$. We first define $\theta : \text{supp}(\psi_1) \to \text{supp}(\psi)$: Let $x \in \text{supp}(\psi_1)$; if $\psi_1(x) \notin E$ then there is a unique $y \in I$ with $\psi(y) = \psi_1(x)$, and in this case we let $\theta(x) = y$; if $\psi_1(x) \in E$ then x is either the left-hand or the right-hand end-point of $\psi_1^{-1}(\{\psi_1(x)\})$, and here we let $\theta(x)$ be the corresponding end-point of $\psi^{-1}(\{\psi_1(x)\})$. We then have $\psi_1(x) = \psi(\theta(x))$ for all $x \in \text{supp}(\psi_1)$, since $\theta(x) \in \psi^{-1}(\{\psi_1(x)\})$ for each $x \in \text{supp}(\psi_1)$, and it is easy to see that θ is strictly increasing and that θ maps $\mathrm{supp}(\psi_1)$ homeomorphically onto $\mathrm{supp}(\psi)$. Also we have $f(\theta(x)) = \theta(f_1(x))$ for all $x \in \mathrm{supp}(\psi_1)$. (If $x \in \mathrm{supp}(\psi_1)$ then

$$\psi(\theta(f_1(x))) = \psi_1(f_1(x)) = g(\psi_1(x)) = g(\psi(\theta(x))) = \psi(f(\theta(x))) ,$$

and thus $\theta(f_1(x)) = f(\theta(x))$ when $\psi(\theta(f_1(x))) \notin E$; it therefore holds for all $x \in \mathrm{supp}(\psi_1)$, because θ is continuous.) We next define θ on $I\text{-}\mathrm{supp}(\psi_1)$: Let $[c,d] = \psi^{-1}(\{\gamma\})$ and $[c_1,d_1] = \psi_1^{-1}(\{\gamma\})$; by assumption there exists a homeomorphism $\sigma : [c_1,d_1] \to [c,d]$ with $\sigma \circ f_1^m = f^m \circ \sigma$, and σ must be increasing (since $\theta(f_1^m(c_1)) = f^m(\theta(c_1))$ and $\theta(c_1) = c$, $\theta(d_1) = d$). Let U_1 be a component of $I\text{-}\mathrm{supp}(\psi_1)$ and let U be the corresponding component of $I\text{-}\mathrm{supp}(\psi)$; we thus have $\psi_1(U) = \{z\} = \psi(U)$ for some $z \in E$. Then if $n = \min\{ k \geq 0 : g^k(z) = \gamma \}$ it follows that f_1^n maps $\overline{U_1}$ homeomorphically onto $[c_1,d_1]$ (since $f_1 \in \mathbf{S}$) and f^n maps $\overline{U}$ homeomorphically onto $[c,d]$ (since $f \in \mathbf{S}$). We define $\theta : \overline{U_1} \to \overline{U}$ by $\theta = (f^n)^{-1} \circ \sigma \circ f_1^n$, and this then gives us a mapping $\theta : I \to I$. It is easily checked that θ is strictly increasing and onto, and hence θ is a homeomorphism; we clearly have $\psi_1 = \psi \circ \theta$. It thus remains to show that $f(\theta(x)) = \theta(f_1(x))$ for each $x \in I\text{-}\mathrm{supp}(\psi_1)$. Let U_1, U and n be as before; if $n \geq 1$ then

$$\theta(f_1(x)) = ((f^{n-1})^{-1} \circ \sigma \circ f_1^{n-1})(f_1(x)) = f(((f^n)^{-1} \circ \sigma \circ f_1^n)(x)) = f(\theta(x))$$

for each $x \in U_1$, and so we are left with the case when $n = 0$, i.e. when $U_1 = (c_1,d_1)$, $U = (c,d)$ and $\theta(x) = \sigma(x)$ for each $x \in U_1$. But here we have

$$\begin{aligned}\theta(f_1(x)) = ((f^{m-1})^{-1} \circ \sigma \circ f_1^{m-1})(f_1(x)) &= (f^{m-1})^{-1}((\sigma \circ f_1^m)(x)) \\ &= (f^{m-1})^{-1}(f^m \circ \sigma)(x) = f(\sigma(x)) = f(\theta(x))\end{aligned}$$

for each $x \in U_1$. □

Theorem 12.3 Each element of $\mathbf{S}_t$ possesses an extension sequence.

Proof Let $f \in \mathbf{S}_t$; if f is essentially transitive then f is an extension sequence for f, and so we can assume that f is not essentially transitive. Then by Theorem 10.12 there exists an essentially transitive mapping $f_0 \in \mathbf{M}_*(I)$, $n \geq 1$ and $[\psi_1,f_1],\ldots,[\psi_n,f_n]$ such that $[\psi_k,f_k]$ is a primary extension of f_{k-1} for $k = 1,\ldots, n$ and $f = f_n$. In fact we must have $f_k \in \mathbf{S}_t$ for each k, since

$(\psi_n\circ\cdots\circ\psi_{k+1},f_k)$ is a reduction of f for $k = 0,\ldots, n-1$, and by Proposition 8.17 the turning point of f_k is periodic for each $k = 0,\ldots, n-1$. Now fix k with $0 < k \le n$; $[\psi_k,f_k]$ is a primary extension of f_{k-1} , thus $\psi_k(I\text{-supp}(\psi_k)) = E$ for some periodic $E \in \mathbf{E}_0(f_{k-1})$ and the f_k-cycle $C = \psi_k^{-1}(E \cap \mathrm{Per}(f_{k-1}))$ is essentially transitive. In particular, the turning point of f_k lies in $\mathrm{int}(C)$, and so if γ is the turning point of f_{k-1} then by Theorem 7.8 (1) $\gamma \in \psi_k(C) \subset E$. Since γ is periodic this means that $E \subset E(\gamma,f_{k-1})$, and hence $E = E(\gamma,f_{k-1})$ because by Lemma 10.6 $\mathbf{E}(\gamma,f_{k-1}) = \{E(\gamma,f_{k-1})\}$; i.e. we have $\psi_k(I\text{-supp}(\psi_k)) = E(\gamma,f_{k-1})$. Let m be the period of γ and let $B = \psi_k^{-1}(\{\gamma\})$; then the restriction of f_k^m to B is essentially transitive (since C is essentially transitive), and if $B = [c,d]$ then $f_k^m(\{c,d\}) \subset \{c,d\}$, because c , $d \in \mathrm{supp}(\psi_k)$. The restriction of f_k^m to B is thus conjugate to an essentially transitive element $h_k \in \mathbf{S}_t$, and this shows that f_k is an h_k extension of f_{k-1} . For each $k = 1,\ldots, n$ we obtain such an $h_k \in \mathbf{S}_t$, and then $f_0, h_1,\ldots, h_n$ is an extension sequence for f . □

Theorem 12.4 Let $f \in \mathbf{S}_t$ and let $h_0,\ldots, h_n$ and $h_0',\ldots, h_m'$ be two extension sequences for f . Then $m = n$ and h_k and h_k' are conjugate for each $k = 0,\ldots, n$.

Proof If f is essentially transitive then it is clear that $m = n = 0$, and so f , h_0 and h_0' are all conjugate; thus we can assume that f is not essentially transitive. Then $\min\{m,n\} \ge 1$, and there exist essentially transitive mappings f_0 , $f_0' \in \mathbf{S}_t$ and $[\psi_1,f_1],\ldots, [\psi_n,f_n], [\psi_1',f_1'],\ldots, [\psi_m',f_m']$ such that f_0 and h_0 are conjugate, f_0' and h_0' are conjugate, $[\psi_k,f_k]$ is an h_k extension of f_{k-1} for each $k = 1,\ldots, n$, $[\psi_k',f_k']$ is an h_k' extension of f_{k-1}' for each $k = 1,\ldots, m$, and f , f_n and f_m' are conjugate. Let $\theta \in \mathbf{V}(I)$ be a homeomorphism with $f_m'\circ\theta = \theta\circ f_n$. We show first that $m = n$, and that there exist unique homeomorphisms $\theta_k \in \mathbf{V}(I)$, $k = 0,\ldots, n-1$, such that $f_k'\circ\theta_k = \theta_k\circ f_k$ for $k = 1,\ldots, n$ and $\theta_{k-1}\circ\psi_k = \psi_k'\circ\theta_k$ for $k = 1,\ldots, n$, where $\theta_n = \theta$. Fix k with $0 \le k \le \min\{m,n\}$ and suppose there exists a homeomorphism $\alpha \in V(I)$ with $f_{m-k}'\circ\alpha = \alpha\circ f_{n-k}$. Let C (resp. C') be the single topologically transitive f_{n-k}-cycle (resp. f_{m-k}'-cycle). Then we must have $\mathrm{supp}(\psi_{n-k}) = \kappa(I\text{-}A(C,f_{n-k}))$ and $\mathrm{supp}(\psi_{m-k}') = \kappa(I\text{-}A(C',f_{m-k}'))$, since by Propositions 10.10 and 12.1 (ψ_{n-k},f_{n-k-1})

(resp. $(\psi'_{m-k}, f'_{m-k-1})$) is a primary reduction of f_{n-k} (resp. of f'_{m-k}). But $\alpha(C) = C'$ and hence $\mathrm{supp}(\psi_{n-k}) = \mathrm{supp}(\psi'_{m-k}\circ\alpha)$. Thus by Proposition 7.1 (3) there exists a unique homeomorphism $\beta \in V(I)$ such that $\psi'_{m-k}\circ\alpha = \beta\circ\psi_{n-k}$, and then also $f'_{m-k-1}\circ\beta = \beta\circ f_{n-k-1}$. Therefore by induction there exists a homeomorphism $\sigma \in V(I)$ such that $f'_{m-p}\circ\sigma = \sigma\circ f_{n-p}$, where $p = \min\{m,n\}$. However, at least one of f_{n-p} and f'_{m-p} is essentially transitive, and so they both must be. This is only possible if $m = n$, and then, again by induction, we obtain the unique homeomorphisms $\theta_{n-1},\ldots, \theta_0 \in V(I)$.

We now show that h_k and h'_k are conjugate for each $k = 0,\ldots, n$. h_0 and h'_0 are conjugate, since $f'_0\circ\theta_0 = \theta_0\circ f_0$, and so fix k with $1 \le k \le n$. Let γ (resp. γ') be the turning point of f_{k-1} (resp. f'_{k-1}), and let $B = \psi_k^{-1}(\{\gamma\})$, $B' = (\psi'_k)^{-1}(\{\gamma'\})$. Then $\theta_k(B) = B'$ and $\theta_k(f_k^q(x)) = (f'_k)^q(\theta_k(x))$ for all $x \in B$, where q is the period of γ . This shows that the restriction of f_k^q to B is conjugate to the restriction of $(f'_k)^q$ to B' , and therefore h_k and h'_k are conjugate. □

Proposition 12.5 u_β is essentially transitive for each $\beta \in (1,2]$.

Proof Suppose first that $\beta \in (\surd 2,2]$. Let C be a u_β-cycle with period m , let B be one of the m components of C and let g be the restriction of f^m to B . Then g is uniformly piecewise linear with slope β^m , and since g has only one turning point we must have $\beta^m \le 2$. This is only possible if $m = 1$, i.e. there are no u_β-cycles with period $m > 1$. Now let $J = [u_\beta^2(\frac{1}{2}), u_\beta(\frac{1}{2})]$; then $u_\beta(J) = J$, and so J is a u_β-cycle with period 1 . Moreover, if K is any u_β-cycle with period 1 then $J \subset K$, since $\frac{1}{2} \in K$. Thus if K is any u_β-cycle with $K \subset J$ then $K = J$, and hence by Lemma 3.7 J is topologically transitive. But it is clear that $I - A(J,u_\beta) \subset \{0,1\}$, and this shows that u_β is essentially transitive. Now suppose that $\beta \in (1,\surd 2]$; then $d = (1+\beta)^{-1}\beta$ is the unique fixed point of u_β in $(\frac{1}{2},1)$. Let $c = 1-d$; we thus have $u_\beta(c) = d$. Then $u_\beta^2([c,d]) \subset [c,d]$, and it is easy to see that the restriction of u_β^2 to $[c,d]$ is conjugate to u_{β^2} . (See the picture on the following page.)

But $f^m(x) \in [c,d]$ for some $m \ge 0$ for each $x \in (0,1)$, and for each $x \in I$ $\{ y \in I : f^n(y) = x$ for some $n \ge 0 \}$ is countable; it thus follows that if u_{β^2}

is essentially transitive then so is u_β. Therefore u_β is essentially transitive for each $\beta \in (1,2]$. □

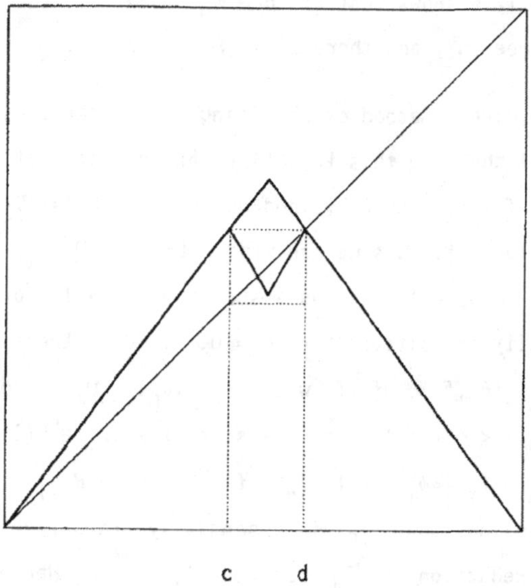

Proposition 12.6 Let

$$P = \{ \beta \in (1,2] : \text{the turning point of } u_\beta \text{ is periodic} \} .$$

Then each element of **P** is an algebraic number, and so in particular **P** is countable.

Proof Let $\beta \in (1,2]$, $x \in [0,1]$ and $n \geq 1$; then $(u_\beta)^n(x)$ is either $\beta(u_\beta)^{n-1}(x)$ or $\beta - \beta(u_\beta)^{n-1}(x)$. It follows by induction on n that for each $n \geq 1$ there exist 2^n polynomials $p_{n,j}$, $1 \leq j \leq 2^n$, each of degree at most n and having coefficients in $\{-1,0,1\}$, such that for every $\beta \in (1,2]$, $x \in [0,1]$ and $n \geq 1$ we have $(u_\beta)^n(x) = p_{n,j}(\beta) + (-1)^\delta \beta^n x$ for some j, $1 \leq j \leq 2^n$, and some $\delta \in \{0,1\}$. Now fix $\beta \in (1,2]$; if the turning point of u_β (i.e. $\frac{1}{2}$) is periodic with period n then $p_{n,j}(\beta) + \frac{1}{2}(-1)^\delta \beta^n = \frac{1}{2}$ for some j, $1 \leq j \leq 2^n$, and some $\delta \in \{0,1\}$, and thus β is a root of some polynomial of degree n having coefficients in the set $\{-2,-1,0,1,2\}$. In particular, β is an algebraic number. □

Remark: For $n \geq 1$ let

$$\mathbf{P}_n = \{ \beta \in (1,2] : \text{the turning point of } u_\beta \text{ is periodic with period } n \} .$$

The proof of Proposition 12.6 shows that if $\beta \in \mathbf{P}_n$ then β is a root of one of 2^{n+1} polynomials of degree n, and therefore $|\mathbf{P}_n| \leq n2^{n+1}$.

We now give a more direct method of obtaining the characteristic sequence of a mapping $f \in \mathbf{S}_t$; we will then use this to define the characteristic sequence for mappings from $\mathbf{S}_r$. Let $f \in \mathbf{S}_t$; then by Theorems 8.4 and 8.14 $\mathbf{D}(f)$ is finite and linearly ordered (by inclusion). Thus we can write $\mathbf{D}(f) = \{D_0, D_1, \ldots, D_n\}$ with $n \geq 0$ and $D_0 \subset D_1 \subset \cdots \subset D_n = I$. Let us assume that $n \geq 1$, or, equivalently, that f is not essentially transitive. For $m = 0, \ldots, n-1$ there exists (by Theorem 7.4) a reduction (θ_m, f_m) of f with $\mathrm{supp}(\theta_m) = D_m$, and since $\mathrm{supp}(\theta_{m-1}) \subset \mathrm{supp}(\theta_m)$, $1 \leq m \leq n-1$, there also exists $\psi_m \in V(I)$ with $\theta_{m-1} = \psi_m \circ \theta_m$ and $\psi_m \circ f_m = f_{m-1} \circ \psi_m$. Put $f_n = f$ and $\psi_n = \theta_{n-1}$. From the results of Section 8 it easily follows that f_0 is essentially transitive and that (ψ_m, f_{m-1}) is a primary reduction of f_m for $m = 1, \ldots, n$. Hence by Proposition 10.10 $[\psi_m, f_m]$ is a primary extension of f_{m-1} for $m = 1, \ldots, n$. Let $1 \leq m \leq n$; then, as in the proof of Theorem 12.3, there exists an essentially transitive mapping $h_m \in \mathbf{S}_t$ such that f_m is an h_m extension of f_{m-1}. There thus also exists a unique $\beta_m \in (1,2]$ such that h_m is conjugate to u_{β_m}. Now by construction $f_0, h_1, \ldots, h_n$ is an extension sequence for f, and therefore $\beta_0, \beta_1, \ldots, \beta_n$ is the characteristic sequence of f, where $\beta_0 \in (1,2]$ is such that f_0 is conjugate to u_{β_0}. If f is essentially transitive then there exists a unique $\beta \in (1,2]$ such that f is conjugate to u_β, and in this case β is the characteristic sequence of f.

The above method of obtaining the characteristic sequence can also be applied to mappings in $\mathbf{S}_r$. We will now use it to define a characteristic sequence for each mapping in $\mathbf{S}_r$, and then show that two mappings in $\mathbf{S}_r$ have the same characteristic sequence if and only if they are conjugate.

Let $f \in \mathbf{S}_r$, and R be the single f-register-shift; suppose first that R is not tame. Then by Theorems 8.10 and 8.14 $\mathbf{D}(f)$ is countably infinite and we can write $\mathbf{D}(f) - \{I\} = \{D_0, D_1, D_2, \ldots \}$ with $D_m \subset D_{m+1}$ for each $m \geq 0$. As before there exists for each $m \geq 0$ a reduction (θ_m, f_m) of f with $\mathrm{supp}(\theta_m) = D_m$, and for $m \geq 1$ there also exists $\psi_m \in V(I)$ with $\theta_{m-1} = \psi_m \circ \theta_m$ and $\psi_m \circ f_m = f_{m-1} \circ \psi_m$. Then, again as before, f_0 is essentially transitive and $[\psi_m, f_m]$ is a primary extension of f_{m-1} for each $m \geq 1$. Let $m \geq 1$; as in the proof of Theorem 12.3 there exists an essentially transitive mapping $h_m \in \mathbf{S}_t$ such that $[\psi_m, f_m]$ is an h_m extension of f_{m-1}, and there thus also exists a unique $\beta_m \in (1,2]$ such that h_m is conjugate to u_{β_m}. Let β_0 be such that f_0 is conjugate to u_{β_0}.

The sequence $\beta_0, \beta_1, \ldots$ which we have constructed depends only on f: To see this let (θ'_m, f'_m) be another reduction of f with $\mathrm{supp}(\theta'_m) = D_m$; and for $m \geq 1$ let $\psi'_m \in V(I)$ be such that $\psi'_m \circ f'_m = f'_{m-1} \circ \psi'_m$. Then by Theorem 7.4 there exists for each $m \geq 0$ a homeomorphism $\alpha_m \in V(I)$ such that $\theta'_m = \alpha_m \circ \theta_m$ and $f'_m \circ \alpha_m = \alpha_m \circ f_m$. For $m \geq 1$ we thus also have

$$\psi'_m \circ \alpha_m \circ \theta_m = \psi'_m \circ \theta'_m = \theta'_{m-1} = \alpha_{m-1} \circ \theta_{m-1} = \alpha_{m-1} \circ \psi_m \circ \theta_m ,$$

and hence $\psi'_m \circ \alpha_m = \alpha_{m-1} \circ \psi_m$, because θ_m is onto. It follows that if $h'_m \in \mathbf{S}_t$ is such that $[\psi'_m, f'_m]$ is an h'_m extension of f'_{m-1} then h'_m and h_m are conjugate, and therefore h'_m and u_{β_m} are conjugate. Finally, f'_0 is conjugate to u_{β_0}, since f'_0 and f_0 are conjugate. We can thus define $\beta_0, \beta_1, \ldots$ to be the characteristic sequence of f. Note that $\beta_m \in \mathbf{P}$ for each $m \geq 0$.

We now consider the case in which the single f-register-shift R is tame. By Theorems 8.10 and 8.14 $\mathbf{D}(f)$ is finite and linearly ordered, and so we can write $\mathbf{D}(f) = \{D_0, D_1, \ldots, D_n\}$ with $n \geq 0$ and $D_0 \subset D_1 \subset \cdots \subset D_n = I$. Assume first that $n \geq 1$, and let (ψ, g) be a reduction of f with $\mathrm{supp}(\psi) = D_{n-1}$. By Lemma 8.15 $g \in \mathbf{S}_t$; hence let $\beta_0, \beta_1, \ldots, \beta_p$ be the characteristic sequence of g. (In fact it is not hard to see that $p = n-1$.) Note that if (ψ', g') is another reduction of f with $\mathrm{supp}(\psi') = D_{n-1}$ then g and g' are conjugate, and so $\beta_0, \beta_1, \ldots, \beta_p$ is also the characteristic sequence of g'. We can thus define $\beta_0, \beta_1, \ldots, \beta_p, 1$ to be the characteristic sequence of f; we have $\beta_k \in \mathbf{P}$ for $k = 0, \ldots, p$.

If $n = 0$ then we define 1 to be the characteristic sequence of f. It will be convenient to denote this set of mappings (i.e. those mappings having 1 as their characteristic sequence) by $\mathbf{S}_r^*$. In fact, if $f \in \mathbf{S}_r$ and R is the single f-register-shift then it is easy to see that $f \in \mathbf{S}_r^*$ if and only if $I - A(R,f)$ is countable.

We have now defined the characteristic sequence for each mapping $f \in S_r$. It is clear that if two mappings in S_r are conjugate then they both have the same characteristic sequence.

Theorem 12.7 If two mappings from S_r have the same characteristic sequence then they are conjugate.

Proof Let f , $f' \in S_r$ have the same characteristic sequence, and suppose first that the single f-register-shift (and thus also the single f′-register-shift) is not tame. Let $\beta_0, \beta_1, \ldots$ be the characteristic sequence of f and f' . Write $D(f) - \{I\} = \{D_0, D_1, \ldots\}$ with $D_m \subset D_{m+1}$ for each $m \geq 0$; for $m \geq 0$ let (θ_m, f_m) be a reduction of f with $\mathrm{supp}(\theta_m) = D_m$, and for $m \geq 1$ let $\psi_m \in V(I)$ be such that $\theta_{m-1} = \psi_m \circ \theta_m$ and $\psi_m \circ f_m = f_{m-1} \circ \psi_m$. Let D'_m , (θ'_m, f'_m) and ψ'_m be the corresponding objects obtained using f' instead of f . Then f_0 and f'_0 are conjugate (since they are both conjugate to u_{β_0}), and for each $m \geq 1$ we have that $[\psi_m, f_m]$ is a u_{β_m} extension of f_{m-1} and $[\psi'_m, f'_m]$ is a u_{β_m} extension of f'_{m-1} . It thus follows from Proposition 12.1 that for each $m \geq 0$ there exists a homeomorphism $\alpha_m \in V(I)$ such that $f'_m \circ \alpha_m = \alpha_m \circ f_m$ and $\psi'_m \circ \alpha_m = \alpha_{m-1} \circ \psi_m$. (A homeomorphism $\alpha_0 \in V(I)$ exists with $f'_0 \circ \alpha_0 = \alpha_0 \circ f_0$ because f_0 and f'_0 are conjugate. Let $m \geq 1$, and suppose a homeomorphism $\alpha_{m-1} \in V(I)$ exists with $f'_{m-1} \circ \alpha_{m-1} = \alpha_{m-1} \circ f_{m-1}$; then it is easy to check that $[\alpha_{m-1} \circ \psi_m, f_m]$ is a u_{β_m} extension of f'_{m-1} , and so by Proposition 12.1 there exists a homeomorphism $\alpha_m \in V(I)$ such that $\alpha_{m-1} \circ \psi_m = \psi'_m \circ \alpha_m$ and $f'_m \circ \alpha_m = \alpha_m \circ f_m$.)

Now for each $m \geq 0$ we define a mapping $\gamma_m : D_m \to D'_m$ as follows: Let $\gamma^\ell_m(x)$ (resp. $\gamma^r_m(x)$) be the smallest (resp. largest) element in the interval $(\theta'_m)^{-1}(\{\alpha_m(\theta_m(x))\})$, (so $\gamma^\ell_m(x) \neq \gamma^r_m(x)$ if and only if this interval is non-trivial), and for $x \in D_m$ let $\gamma_m(x) = \gamma^\ell_m(x)$ (resp. $\gamma_m(x) = \gamma^r_m(x)$) if x is the smallest (resp. the largest) element of the interval $(\theta_m)^{-1}(\{\theta_m(x)\})$. (This definition makes sense because $(\theta_m)^{-1}(\{y\})$ is a non-trivial interval if and only if $(\theta'_m)^{-1}(\{\alpha_m(y)\})$ is a non-trivial interval. In fact $(\theta_m)^{-1}(\{y\})$ (resp. $(\theta'_m)^{-1}(\{z\})$) is non-trivial if and only if $(f_m)^n(y) = \sigma_m$ for some $n \geq 0$ (resp.

$(f'_m)^n(z) = \sigma'_m$ for some $n \geq 0$), where σ_m (resp. σ'_m) is the turning point of f_m (resp. of f'_m).) It is easily seen that γ_m maps D_m bijectively onto D'_m , and γ_m is strictly increasing. Also $\gamma_m(f(x)) = f'(\gamma_m(x))$ for all $x \in D_m$, (since $\alpha_m(\theta_m(f(x))) = f'_m(\alpha_m(\theta_m(x)))$ and $f'_m \circ \theta'_m = \theta'_m \circ f'$); moreover, the mappings $\{\gamma_m\}_{m\geq 0}$ are compatible, in that $\gamma_{m+1}(x) = \gamma_m(x)$ for all $x \in D_m$, $m \geq 1$ (because $\alpha_m(\theta_m(x)) = \psi'_{m+1}(\alpha_{m+1}(\theta_{m+1}(x)))$ and $\theta'_m = \psi'_{m+1} \circ \theta'_{m+1}$). Put $D = \bigcup_{m\geq 0} D_m$ and $D' = \bigcup_{m\geq 0} D'_m$; we can thus define a mapping $\gamma : D \to D'$ by letting $\gamma(x) = \gamma_m(x)$ for each $x \in D_m$. We immediately have that γ maps D bijectively onto D' , γ is strictly increasing, and $\gamma(f(x)) = f'(\gamma(x))$ for all $x \in D$.

Lemma 12.8 D and D' are dense subsets of I .

Proof Suppose D is not dense; then by Lemma 8.2 $U = \mathrm{int}(I-D)$ is a non-empty f-almost-invariant open set and, since $f \in \mathbf{M}_0(I)$, there exists a periodic component J of U containing the turning point of f . Let q be the period of J and for $k = 0,\ldots, q-1$ let J_k be the component of U such that $f^k(J) \subset \overline{J_k}$. Then $C = \bigcup_{k=0}^{q-1} \overline{J_k}$ is an f-cycle, and $\mathrm{int}(C) \subset U \subset I-D_m$ for each $m \geq 0$. Let $\{K_n\}_{n\geq 1}$ be a generator for the single f-register-shift R , and for $n \geq 1$ let $E_n = \kappa(I-A(K_n,f))$. For each $n \geq 1$ we thus have either $E_n = \varnothing$ or $E_n = D_m$ for some $m \geq 0$, and hence $C - A(K_n,f)$ is countable. Therefore $C - A(R,f)$ is countable, which contradicts the assumption that R is not tame (because it is clear that $R \subset C$). This shows that D is dense, and in the same way D' is also dense. □

Lemma 12.9 Let E and E' be dense subsets of I , and suppose $\sigma : E \to E'$ is strictly increasing and maps E onto E' . Then there exists a unique homeomorphism $\tau \in V(I)$ with $\tau(x) = \sigma(x)$ for all $x \in E$.

Proof Easy exercise. □

By Lemmas 12.8 and 12.9 there exists a homeomorphism $\pi \in V(I)$ such that $\pi(x) = \gamma(x)$ for all $x \in D$. Then by the continuity of f, f' and π we have $\pi(f(x)) = f'(\pi(x))$ for all $x \in I$, and hence f and f' are conjugate.

It remains to consider the case when the single f-register-shift (and thus also the single f′-register-shift) is tame. Assume first that $f \notin \mathbf{S}_r^*$, and so the characteristic sequence of f and f′ has the form $\beta_0, \ldots, \beta_p, 1$ for some $p \geq 0$. Let D (resp. D′) be the largest element in $\mathbf{D}(f) - \{I\}$ (resp. $\mathbf{D}(f') - \{I\}$), and let (ψ, g) (resp. (ψ', g')) be a reduction of f with $\mathrm{supp}(\psi) = D$ (resp. a reduction of f′ with $\mathrm{supp}(\psi') = D'$). Then $\beta_0, \ldots, \beta_p$ is the characteristic sequence of both g and g′ , and hence by Proposition 12.2 g and g′ are conjugate. Now as in the proof of Theorem 12.3 there exist $q, q' \in \mathbf{S}$ such that $[\psi, f]$ is a q extension of g and $[\psi', f']$ is a q′ extension of g′ , and if q and q′ are conjugate then by Proposition 12.2 f and f′ will be conjugate. But it is easy to see that we must have $q, q' \in \mathbf{S}_r^*$; thus the proof of Theorem 12.7 will be complete if we can show that any two elements in $\mathbf{S}_r^*$ are conjugate, and this fact is the content of the result which follows. □

Proposition 12.10 Any two elements in $\mathbf{S}_r^*$ are conjugate.

Proof Let $f, g \in \mathbf{S}_r^*$. We will define two sequences $\{M_n\}_{n\geq 1}$ and $\{N_n\}_{n\geq 1}$ of subsets of I , and for each $n \geq 1$ a mapping $\psi_n : M_n \to N_n$ such that:

(12.3) $M_n \subset M_{n+1}$ and $N_n \subset N_{n+1}$ for each $n \geq 1$.

(12.4) M_n is f-invariant and N_n is g-invariant for each $n \geq 1$.

(12.5) For each $n \geq 1$ we have ψ_n is strictly increasing and ψ_n maps M_n onto N_n .

(12.6) $\psi_n(f(x)) = g(\psi_n(x))$ for all $x \in M_n$, $n \geq 1$.

(12.7) $\psi_n(x) = \psi_{n+1}(x)$ for all $x \in M_n$, $n \geq 1$.

(12.8) $\bigcup_{n\geq 1} M_n$ and $\bigcup_{n\geq 1} N_n$ are both dense subsets of I .

If we have M_n , N_n and ψ_n , $n \geq 1$, satisfying these conditions then it follows immediately that there exists a homeomorphism $\theta \in \mathbf{V}(I)$ with $\theta \circ f = g \circ \theta$: Put $M = \bigcup_{n\geq 1} M_n$ and $N = \bigcup_{n\geq 1} N_n$; from (12.3) and (12.7) we can define $\psi : M \to N$ by letting $\psi(x) = \psi_n(x)$ for $x \in M_n$. Thus ψ is strictly increasing and maps M onto N ; moreover, M and N are both dense in I . Hence by Lemma 12.9 there exists a unique homeomorphism $\theta \in \mathbf{V}(I)$ with $\theta(x) = \psi(x)$ for each $x \in M$, and we then have $\theta \circ f = g \circ \theta$, because $\psi(f(x)) = g(\psi(x))$ for all $x \in M$.

We now start the construction of the sets M_n , N_n and the mappings ψ_n . Let γ be the turning point of f . Since $Z(f) = \emptyset$ we have $f(x) > x$ for all $x \in (0,\gamma]$, and so f has a unique fixed point $\beta \in (\gamma,1)$; there thus also exists a unique $\alpha \in (0,\gamma)$ with $f(\alpha) = \beta$. Now we must have $f^2([\alpha,\beta]) \subset [\alpha,\beta]$ or, equivalently, $f^2(\gamma) \geq \gamma$. (To see this, let $W = \{ x \in I : f^2(x) = \beta$ for some $n \geq 0 \}$ and $\{K_n\}_{n\geq 1}$ be a generator for the single f-register-shift R . Then $W \cap A(K_n,f) = \emptyset$ and hence also $\overline{W} \cap A(K_n,f) = \emptyset$ for each $n > 1$, since $\beta \in Fix(f)$ and $per(K_2) > 1$. Therefore $\overline{W} \subset I-A(R,f)$, which implies that $\overline{W}$ is countable, because $f \in S_r^*$. But if $f^2(\gamma) < \gamma$ then $\overline{W}$ is uncountable. (See Lemma 13.7 for a proof of this, if necessary.))

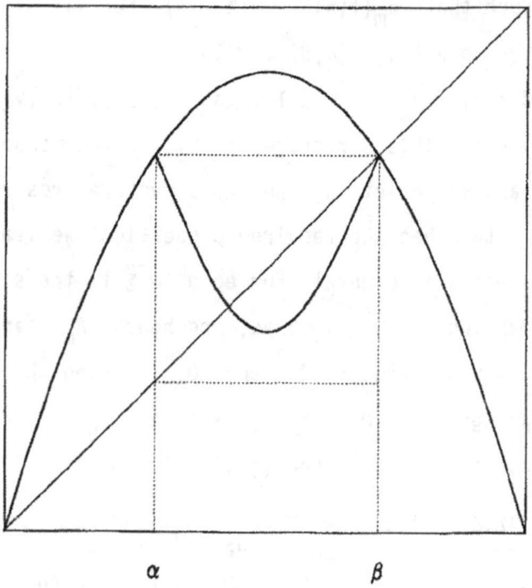

We have $f^2([\alpha,\beta]) \subset [\alpha,\beta]$, and so we can consider $h \in M([\alpha,\beta])$, where h is the restriction of f^2 to $[\alpha,\beta]$. It is clear that h is conjugate (via a linear change of variables) to an element $f' \in \mathbf{S}$, and in fact $f' \in S_r^*$ (since $\{ x \in I : f^n(x) \in [\alpha,\beta]$ for some $n \geq 0 \} = (0,1)$). We can thus repeat the above construction using f' instead of f , and in fact we can repeat the construction indefinitely, because at each stage we obtain a new element of S_r^* . This procedure shows, in particular, that for each $n \geq 1$ f has exactly one periodic orbit with period 2^n . (For example, f^2 has a unique fixed point z in (α,β) , and then

$\{z,f(z)\}$ is the single periodic orbit of f with period 2 .) Moreover, if q is not of the form 2^n for some $n \geq 0$ then there are no periodic points with period q . (This also follows from Theorem 9.7.) For each $m \geq 1$ let β_m be one of the points from the periodic orbit with period 2^m , and put $P_m = \{ x \in I : f^k(x) = \beta_m$ for some $k \geq 0 \}$. (Note that P_m is independent of the choice of β_m .) Let $M_n = \bigcup_{m=1}^{n} P_m$; thus for each $n \geq 1$ we have $M_n \subset M_{n+1}$ and M_n is f-invariant. In the same way we define $N_n = \bigcup_{m=1}^{n} Q_m$, where $Q_m = \{ x \in I : g^k(x) = \nu_m$ for some $k \geq 0 \}$ and ν_m is one of the points from the single periodic orbit of g with period 2^m . Clearly we have (12.3) and (12.4) .

Now for each $m \geq 1$ there is a "canonical" bijection $\pi_m : P_m \to Q_m$, which is strictly increasing and such that $\pi_m(f(x)) = g(\pi_m(x))$ for all $x \in P_m$. (We have $P_1 = \{ u_n : n \geq 1 \} \cup \{ v_n : n \geq 1 \} \cup \{\alpha,\beta\}$, with $0 < \cdots < u_2 < u_1 < \alpha < \beta < v_1 < v_2 < \cdots < 1$, and $f(u_{p+1}) = f(v_{p+1}) = u_p$ for $p \geq 1$ and $f(u_1) = f(v_1) = \alpha$. This structure of P_1 is independent of which mapping in $\mathbf{S}_r^*$ is used, and so we let π_1 be the structure-preserving mapping between P_1 and Q_1 . π_1 then has the required properties. We leave it for the reader to check that this works in general: For each $m \geq 1$ the structure of P_m is independent of which mapping in $\mathbf{S}_r^*$ is used, and hence π_m can be defined as the structure-preserving mapping between P_m and Q_m .) Using the mappings π_m , $m \geq 1$, we define the mappings $\psi_n : M_n \to N_n$ by letting $\psi_n(x) = \pi_m(x)$ for $x \in P_m$, $m \leq n$. It is easy to see that then (12.5), (12.6) and (12.7) hold.

It remains to show that (12.8) holds. Let $M = \bigcup_{n\geq 1} M_n$ and put $U = \mathrm{int}(I-M)$. By Lemma 8.2 U is f-almost-invariant, because $I-M$ is f-invariant. But we have seen that $\mathrm{Per}(f) \subset M$, and thus $U \cap \mathrm{Per}(f) = \varnothing$. Therefore $U = \varnothing$, since otherwise we would have $U \cap Z(f) \neq \varnothing$. (If $f \in \mathbf{M}(I)$ and U is a non-empty f-almost-invariant open subset of I with $U \cap \mathrm{Per}(f) = \varnothing$ then it is easy to check that $U \cap Z(f) \neq \varnothing$.) This shows that M is dense, and in the same way $\bigcup_{n\geq 1} N_n$ is also dense in I . □

Let $f \in \mathbf{S}_r$; then the characteristic sequence of f has one of the following forms:

(1) 1 .

(2) $\beta_0, \ldots, \beta_p$, 1 with $p \geq 0$ and $\beta_k \in \mathbf{P}$ for $k = 0, \ldots, p$.

(3) $\beta_0, \beta_1, \beta_2, \ldots$ with $\beta_m \in \mathbf{P}$ for each $m \geq 0$.

We end this section by showing that, conversely, each such sequence is the characteristic sequence of some element in $\mathbf{S}_r$.

Proposition 12.11 (1) 1 is the characteristic sequence of some element in $\mathbf{S}_r$, i.e. $\mathbf{S}_r^* \neq \emptyset$.

(2) Let $p \geq 0$ and $\beta_0, \ldots, \beta_p \in \mathbf{P}$; then $\beta_0, \ldots, \beta_p$, 1 is the characteristic sequence of some element in $\mathbf{S}_r$.

(3) For each $m \geq 0$ let $\beta_m \in \mathbf{P}$; then $\beta_0, \beta_1, \beta_2, \ldots$ is the characteristic sequence of some element in $\mathbf{S}_r$.

Proof (1): We have already constructed an element of $\mathbf{S}_r^*$ in Section 5.

(2): By Proposition 12.1 $\beta_0, \ldots, \beta_p$ is the characteristic sequence of some element $g \in \mathbf{S}_t$, and the turning point of g is periodic, because $\beta_p \in \mathbf{P}$. Let $q \in \mathbf{S}_r^*$; then, again using Proposition 12.1 there exists a q extension $[\psi, f]$ of g . It is easy to see that $f \in \mathbf{S}_r$ and $\beta_0, \ldots, \beta_p$, 1 is the characteristic sequence of f .

(3): Let $g \in \mathbf{S}_t$ with periodic turning point γ and let $q \in \mathbf{S}$; then in Proposition 12.1 we constructed a q extension $[\psi, f]$ of g . Note that in this construction:

(i) ψ was chosen independently of q .

(ii) If $B = \psi^{-1}(\{\gamma\})$ then the definition of f on $I-B$ did not depend on q .

Thus q only plays a rôle in the definition of f on B . In fact, this construction in the proof of Proposition 12.1 can be formalized as follows: Let $g \in \mathbf{S}_t$ with periodic turning point γ , and choose $\psi \in \mathbf{V}(I)$ with $\psi(I\text{-supp}(\psi)) = E(\gamma, g)$; let $B = \psi^{-1}(\{\gamma\})$ and $J = \psi^{-1}(\{g(\gamma)\})$. Then there exists a mapping $w : I-B \to I$ and homeomorphisms $s : B \to I$ and $t : I \to J$ such that $[\psi, w_q]$ is a q extension of g for each $q \in \mathbf{S}$, where

$$w_q(x) = \begin{cases} w(x) & \text{if } x \in I-B\ , \\ (t \circ q \circ s)(x) & \text{if } x \in B\ . \end{cases}$$

We call $\{\psi, w, s, t\}$ an **extension kit** for g. For $n \geq 0$ let $\{\psi_n, w_n, s_n, t_n\}$ be an extension kit for u_{β_n}. Put $B_n = (\psi_n)^{-1}(\{\frac{1}{2}\})$, $J_n = (\psi_n)^{-1}(\{u_{\beta_n}(\frac{1}{2})\})$ (noting that $\frac{1}{2}$ is the turning point of u_{β_n}), so $w_n : I-B_n \to I$, $s_n : B_n \to I$ and $t_n : I \to J_n$. We define a sequence of intervals $\{K_n\}_{n\geq 0}$ by $K_0 = I$, $K_1 = B_0$ and (for $n \geq 1$) $K_{n+1} = (s_0)^{-1} \circ (s_1)^{-1} \circ \cdots \circ (s_{n-1})^{-1}(B_n)$; this sequence is decreasing because $(s_n)^{-1}(B_{n+1}) \subset B_n$ for each $n \geq 0$. We also define for each $n \geq 1$ a mapping $v_n : K_n - K_{n+1} \to I$ by $v_n = t_0 \circ t_1 \circ \cdots \circ t_{n-1} \circ w_n \circ s_{n-1} \circ \cdots \circ s_1 \circ s_0$, and put $v_0 = w_0$. Now if the mappings $\{\psi_n\}_{n\geq 1}$ are chosen carefully enough then we will have $\lim_{n\to\infty} |K_n| = 0$. (This just amounts to choosing $|B_n|$ small enough for each $n \geq 0$.) We suppose then that this has been done, and let γ be the single point in $\bigcap_{n\geq 0} K_n$ (so $\gamma \in (0,1)$). Define $f : I-\{\gamma\} \to I$ by letting $f(x) = v_n(x)$ for $x \in K_n - K_{n+1}$. It is easy to see that f is continuous, and that f is strictly increasing on $[0,\gamma)$ and strictly decreasing on $(\gamma,1]$. Moreover, $\lim_{x\uparrow\gamma} f(x) = \lim_{x\downarrow\gamma} f(x)$, and thus if we define $f(\gamma) = \lim_{x\uparrow\gamma} f(x)$ then $f \in \mathbf{M}(I)$ and γ is the single turning point of f. We also have $f(0) = f(1) = 0$, and therefore $f \in \mathbf{S}$, provided $Z(f) = \emptyset$. It is now left to the reader to show that f is essentially the mapping with the required characteristic sequence. More precisely, we have: Either $Z(f) = \emptyset$, in which case $f \in \mathbf{S}_r$ and $\beta_0, \beta_1, \beta_2, \ldots$ is the characteristic sequence of f, or $Z(f) \neq \emptyset$, and in this case $I-Z(f)$ is uncountable, and if (ψ, g) is a reduction of f with $\mathrm{supp}(\psi) = \kappa(I-Z(f))$ then $g \in \mathbf{S}_r$ and $\beta_0, \beta_1, \beta_2, \ldots$ is the characteristic sequence of g. □

13. SOME MISCELLANEOUS RESULTS FROM REAL ANALYSIS

In this section we prove some of the results from real analysis which we have used in the previous sections. All of these results are standard, or at least fairly standard, but probably some of them do not occur in a typical introductory real analysis text. As before, I denotes the closed interval $[a,b]$.

Let B , $A \subset I$ with $B \subset A$; we say that B is dense in A if for each $x \in A$ and each $\varepsilon > 0$ there exists $y \in B$ with $|y-x| < \varepsilon$. The first result of this section is a simple version of the Baire Category Theorem.

Theorem 13.1 Let D be a non-empty closed subset of I , and for each $n \geq 1$ let U_n be an open subset of I such that $U_n \cap D$ is dense in D . Then $(\bigcap_{n\geq 1} U_n) \cap D$ is also dense in D . In particular, if U_n is a dense open subset of I for each $n \geq 1$ then $\bigcap_{n\geq 1} U_n$ is also a dense subset of I .

Proof Let $x \in D$ and $\varepsilon > 0$; we must show there exists an element $y \in (\bigcap_{n\geq 1} U_n) \cap D$ with $|y-x| < \varepsilon$. To this end we inductively define a decreasing sequence of non-trivial closed intervals $\{J_n\}_{n\geq 1}$ such that $J_n \subset U_n$ and $\text{int}(J_n) \cap D \neq \varnothing$ for each $n \geq 1$, and also so that $|z-x| < \varepsilon$ for all $z \in J_1$. We first find an interval J_1 with the required properties: $U_1 \cap D$ is dense in D , and hence there exists $u \in U_1 \cap D$ with $|u-x| < \varepsilon$; we can thus let J_1 be a suitable interval with $u \in \text{int}(J_1)$. Next let $n \geq 1$, and suppose that the intervals $J_1, \ldots, J_n$ have already been constructed. Then, since $\text{int}(J_n) \cap D \neq \varnothing$ and $U_{n+1} \cap D$ is dense in D , we have $\text{int}(J_n) \cap U_{n+1} \cap D \neq \varnothing$, and therefore there exists a non-trivial closed interval J_{n+1} with $J_{n+1} \subset J_n \cap U_{n+1}$ and $\text{int}(J_{n+1}) \cap D \neq \varnothing$. This shows that a sequence of intervals $\{J_n\}_{n\geq 1}$ with the required properties does exist. But now it is clear that $(\bigcap_{n\geq 1} J_n) \cap D \neq \varnothing$, and if $y \in (\bigcap_{n\geq 1} J_n) \cap D$ then $y \in (\bigcap_{n\geq 1} U_n) \cap D$ and $|y-x| < \varepsilon$. □

Lemma 13.2 Let D be a non-empty countable closed subset of I . Then D contains an isolated point.

Proof This is clear if D is finite, so suppose D is infinite and let $\{x_n\}_{n\ge1}$ be an enumeration of the elements of D. For each $n \ge 1$ let $U_n = I - \{x_n\}$; then U_n is an open subset of I and $(\bigcap_{n\ge1} U_n) \cap D = \emptyset$. Hence by Theorem 13.1 there must exist $m \ge 1$ such that $U_m \cap D$ is not dense in D. But $U_m \cap D = D - \{x_m\}$, and $D - \{x_m\}$ being not dense in D just means that x_m is an isolated point of D. □

Lemma 13.3 Let D be a non-empty closed subset of I, and let $f : D \to D$ be a continuous bijection. Then f is a homeomorphism (i.e. f^{-1} is also continuous).

Proof Let F be a closed subset of D ; then F is compact, and hence $f(F)$ is compact; thus $f(F)$ is a closed subset of D. Therefore, since f is a bijection, we have that $(f^{-1})^{-1}(F)$ is a closed subset of D for each closed subset F of D. This shows that f^{-1} is continuous. □

Lemma 13.4 Let $\{a_n\}_{n\ge1}$ be a sequence of real numbers such that $a_{m+n} \le a_m + a_n$ for all $m, n \ge 1$. Then $\lim_{n\to\infty} \frac{1}{n} a_n = \inf_{n\ge1} \frac{1}{n} a_n$. (However, it is possible that $\inf_{n\ge1} \frac{1}{n} a_n = -\infty$.)

Proof First note that $a_{km} \le ka_m$ for all $k, m \ge 1$. Let $m \ge 1$, and put $b_m = \max\{a_1,\dots,a_m\}$. Now if $n > m$ then we can write $n = km + \ell$ with $k \ge 1$ and $1 \le \ell \le m$, and we then have

$$\frac{1}{n} a_n = \frac{1}{n} a_{km+\ell} \le \frac{1}{n} a_{km} + \frac{1}{n} a_\ell \le \frac{k}{n} a_m + \frac{1}{n} b_m \le \frac{1}{m} a_m + \frac{1}{n} b_m .$$

It follows that $\limsup_{n\to\infty} \frac{1}{n} a_n \le \frac{1}{m} a_m$ for each $m \ge 1$, and so $\limsup_{n\to\infty} \frac{1}{n} a_n \le \inf_{n\ge1} \frac{1}{n} a_n$. But clearly $\inf_{n\ge1} \frac{1}{n} a_n \le \liminf_{n\to\infty} \frac{1}{n} a_n$, and therefore $\lim_{n\to\infty} \frac{1}{n} a_n = \inf_{n\ge1} \frac{1}{n} a_n$. □

Proposition 13.5 Let D be a non-empty perfect subset of I. Then there exists $\psi \in V(I)$ with $\text{supp}(\psi) = D$.

Proof We first consider the case when D is nowhere dense. Then, since D is perfect, $I-D$ has infinitely many connected components; thus let $\{U_n\}_{n\geq 1}$ be the components of $I-D$. Put $U_\ell = (-\infty,a)$, $U_r = (b,\infty)$, and define a mapping $\tau : \{\ell,r,1,2,\ldots\} \to I$ inductively as follows: $\tau(\ell) = a$, $\tau(r) = b$, $\tau(1) = \frac{1}{2}(a+b)$; now let $n \geq 1$, and suppose $\tau(k)$, $k = 1,\ldots, n$, have already been defined. Then there exist unique elements $p, q \in \{\ell,r,1,\ldots,n\}$ such that U_{n+1} lies between U_p and U_q , but U_m does not lie between U_p and U_q for each $m \in \{\ell,r,1,\ldots,n\}$; we put $\tau(n+1) = \frac{1}{2}(\tau(p)+\tau(q))$. Note that the mapping τ is injective. Using τ we can define a mapping $\alpha : I-D \to I$ by letting $\alpha(x) = \tau(n)$ for each $x \in U_n$, $n \geq 1$. It is easily checked that α is increasing. If $1 \leq m < n$ then, because D is perfect and nowhere dense, there exists $p > n$ such that U_p lies between U_m and U_n , and from this it follows that $\alpha(I-D)$ is a dense subset of I . Therefore, since $I-D$ and $\alpha(I-D)$ are both dense in I and α is increasing, there exists a unique $\psi \in V(I)$ such that $\psi(x) = \alpha(x)$ for each $x \in I-D$. By construction ψ is constant on each component of $I-D$, and so we clearly have $\mathrm{supp}(\psi) \subset D$. Conversely, let $x \in D$ and $V \subset I$ be a neighbourhood of x in I ; then $\{ n \geq 1 : U_n \cap V \neq \emptyset \}$ contains more then one element, and hence ψ is not constant on V (since τ is injective); i.e. $x \in \mathrm{supp}(\psi)$. This shows that $\mathrm{supp}(\psi) = D$.

We now consider the general case (when D is not necessarily nowhere dense). Note that the sets $\mathrm{int}(D)$ and $\mathrm{int}(I-\mathrm{int}(D))$ are disjoint, and their union is dense in I . Let $\{V_\beta\}_{\beta\in M}$ be the connected components of $\mathrm{int}(D)$ and $\{W_\gamma\}_{\gamma\in N}$ the connected components of $\mathrm{int}(I-\mathrm{int}(D))$; also let $N' = \{ \gamma \in N : W_\gamma \cap D \neq \emptyset \}$. If $\gamma \in N'$ then $D_\gamma = \overline{W_\gamma \cap D}$ is a non-empty, perfect, nowhere dense subset of $\overline{W_\gamma} = [a_\gamma,b_\gamma]$, and hence by the first part of the proof there exists $\alpha_\gamma \in V([a_\gamma,b_\gamma])$ with $\mathrm{supp}(\alpha_\gamma) = D_\gamma$. We define $\sigma_\gamma : I \to R$ by

$$\sigma_\gamma(x) = \begin{cases} 0 & \text{if } x \in [a,a_\gamma] , \\ \alpha_\gamma(x)-a_\gamma & \text{if } x \in [a_\gamma,b_\gamma] , \\ b_\gamma-a_\gamma & \text{if } x \in [b_\gamma,b] . \end{cases}$$

Then σ_γ is increasing, continuous and $\mathrm{supp}(\sigma_\gamma) = D_\gamma$. For $\beta \in M$ we define $\tau_\beta : I \to R$ by

$$\tau_\beta(x) = \begin{cases} 0 & \text{if } x \in [a,c_\beta] , \\ x-c_\beta & \text{if } x \in [c_\beta,d_\beta] , \\ d_\beta-c_\beta & \text{if } x \in [d_\beta,b] , \end{cases}$$

where $[c_\beta,d_\beta] = \overline{V_\beta}$; τ_β is increasing, continuous and $\mathrm{supp}(\tau_\beta) = \overline{V_\beta}$. Now the sums $\sum_{\gamma\in N'} \sigma_\gamma$ and $\sum_{\beta\in M} \tau_\beta$ converge uniformly, and so we can define an increasing continuous mapping $\pi : I \to R$ by letting $\pi(x) = \sum_{\gamma\in N'} \sigma_\gamma(x) + \sum_{\beta\in M} \tau_\beta(x)$. It is then easily checked that $\mathrm{supp}(\pi) = D$, and thus we obtain a mapping $\psi \in V(I)$ with $\mathrm{supp}(\psi) = D$ by letting $\psi(x) = a + (b-a)(\pi(x)-\pi(a))/(\pi(b)-\pi(a))$. □

Proposition 13.6 Let $E \subset I$ be countable. Then there exists $\psi \in V(I)$ with $\psi(I-\mathrm{supp}(\psi)) = E$.

Proof Choose a sequence $\{x_n\}_{n\geq 1}$ from (a,b) with the properties:

(i) $x_n \neq x_m$ for all $n \neq m$.

(ii) For each $\varepsilon > 0$ there exists $n \geq 1$ with $\bigcup_{k=1}^{n} B(x_k,\varepsilon) \supset I$, (where $B(x,\varepsilon) = \{ y \in R : |y-x| < \varepsilon \}$).

(iii) For each finite subset F of E there exists $n \geq 1$ such that $F \subset \{a,b,x_1,\dots,x_n\}$.

For $n \geq 1$ let ℓ_n be the largest element in the set $\{a,x_1,\dots,x_{n-1}\} \cap [a,x_n)$ and r_n be the smallest element in the set $\{x_1,\dots,x_{n-1},b\} \cap (x_n,b]$. Thus $\ell_n, r_n \in \{a,x_1,\dots,x_{n-1},b\}$, $\ell_n < x_n < r_n$ and $(\ell_n,r_n) \cap \{a,x_1,\dots,x_{n-1},b\} = \emptyset$. We inductively construct a sequence $\{\psi_n\}_{n\geq 0}$ from $V(I)$ as follows: First choose $a \leq a' < b' \leq b$ with $a' > a$ if and only if $a \in E$ and $b' < b$ if and only if $b \in E$. Let $\psi_0(x) = a$ for $x \in [a,a']$, $\psi_0(x) = b$ for $x \in [b',b]$ and ψ_0 be linear on $[a',b']$. Now let $n \geq 1$. If $x_n \notin E$ then put $\psi_n = \psi_{n-1}$. If $x_n \in E$ then define ψ_n to be equal to ψ_{n-1} except on the interval $M_n = \psi_{n-1}^{-1}((\ell_n,r_n))$; on $\overline{M_n} = [a_n,b_n]$ we define ψ_n by choosing $a_n < \alpha_n < \beta_n < b_n$ and letting ψ_n be linear on $[a_n,\alpha_n]$ and $[\beta_n,b_n]$ and equal to the constant x_n on $[\alpha_n,\beta_n]$. (See the picture on the next page.)

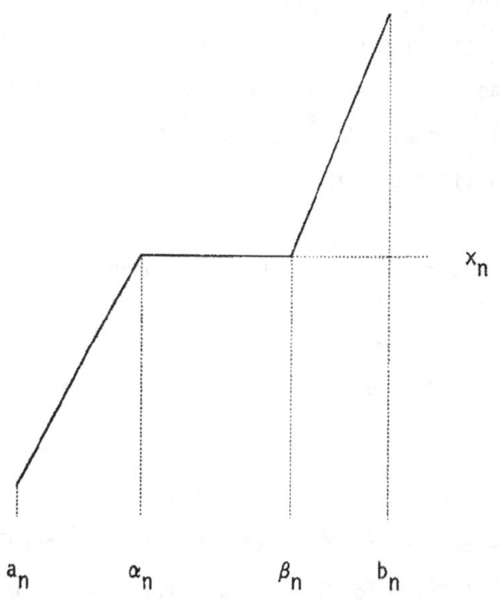

(It is easy to see that ψ_{n-1} is linear on $\overline{M_n}$.) Each ψ_n is clearly piecewise linear. For $n \geq 1$ let ε_n be the length of the largest interval $J \subset I$ with $J \cap \{a,x_1,\dots,x_n,b\} = \emptyset$. By *(ii)* we have $\lim_{n\to\infty} \varepsilon_n = 0$. We next show that if $1 \leq n < m$ then

$$(13.1) \qquad |\psi_n(x) - \psi_m(x)| < \varepsilon_n \quad \text{for all } x \in I .$$

By construction we have $\psi_m = \psi_n$ on $A_n = \psi_n^{-1}(\{a,x_1,\dots,x_n,b\})$. Let $x \in I - A_n$, let u be the largest element in $[a,x] \cap A_n$ and v be the smallest element in $[x,b] \cap A_n$; thus $u < x < v$. We have
$\psi_n(x) - \psi_m(x) < \psi_n(v) - \psi_m(u) = \psi_n(v) - \psi_n(u)$, and also
$\psi_m(x) - \psi_n(x) < \psi_m(v) - \psi_n(u) = \psi_n(v) - \psi_n(u)$; i.e. we have
$|\psi_n(x) - \psi_m(x)| < \psi_n(v) - \psi_n(u)$. But $\psi_n(v) - \psi_n(u) \leq \varepsilon_n$, because $\psi_n((u,v))$ is an interval with $\psi_n((u,v)) \cap \{a,x_1,\dots,x_n,b\} = \emptyset$, and this gives us that (13.1) holds for all $1 \leq n < m$. We can thus define $\psi(x) = \lim_{n\to\infty} \psi_n(x)$, and obtain an element $\psi \in V(I)$. Let $x \in E$; by *(iii)* we have $x = x_m$ for some $m \geq 1$. Then $J = \psi_m^{-1}(\{x\})$ is a non-trivial closed interval and $\psi_n^{-1}(\{x\}) = J$ for all $n \geq m$. Hence $\psi(J) = \{x\}$, and so $x \in \psi(I\text{-supp}(\psi))$; i.e. we have $E \subset \psi(I\text{-supp}(\psi))$.

Finally, let $x \in \psi(\text{I-supp}(\psi))$, and put $\psi^{-1}(\{x\}) = [c,d]$ (so $d > c$). By (13.1) we can find $n \geq 1$ such that $\psi_n(d) - \psi_n(c) < d-c$. But on each interval where it is linear the slope of ψ_n is either 0 or at least 1. There thus exists $z \in E \cap \{a,x_1,\ldots,x_n,b\}$ and $c \leq \alpha < \beta \leq d$ so that $\psi_n([\alpha,\beta]) = \{z\}$. However, we then have $\psi_k([\alpha,\beta]) = \{z\}$ for all $k \geq n$, and hence $\psi([\alpha,\beta]) = \{z\}$. Therefore $x = z \in E$, i.e. $\psi(\text{I-supp}(\psi)) \subset E$. □

Lemma 13.7 Let $g \in M(I)$, $a < \alpha < y < \beta \leq b$, and suppose that $g(\alpha) = g(\beta) = \beta$ and $g(y) < \alpha$; put

$$W = \{ x \in [\alpha,\beta] : g^n(x) = \beta \text{ for some } n \geq 0 \}.$$

Then $\overline{W}$ is uncountable.

Proof For each $m \geq 1$ we construct 2^m disjoint closed intervals $I_k^{(m)}$, $1 \leq k \leq 2^m$, such that $g^m(I_k^{(m)}) \supset [\alpha,\beta]$ for $k = 1,\ldots,2^m$, $m \geq 1$, and so that $I_{2k-1}^{(m+1)} \cup I_{2k}^{(m+1)} \subset I_k^{(m)}$ for $k = 1,\ldots,2^m$, $m \geq 0$, where $I_1^{(0)} = [\alpha,\beta]$. First choose $\sigma \in (\alpha,y)$, $\tau \in (y,\beta)$ with $g(\sigma) = g(\tau) = \alpha$ and put $I_1^{(1)} = [\alpha,\sigma]$ and $I_2^{(1)} = [\tau,\beta]$. Now let $m \geq 1$, and suppose disjoint closed intervals $I_k^{(m)}$, $1 \leq k \leq 2^m$, have been constructed with $g^m(I_k^{(m)}) \supset [\alpha,\beta]$ for each $k = 1,\ldots,2^m$. Fix k, $1 \leq k \leq 2^m$; then we can find elements $s, t, u, v \in I_k^{(m)}$ with either $s < t < u < v$ or $v < u < t < s$ such that $g^m(s) = \alpha$, $g^m(t) = \sigma$, $g^m(u) = \tau$ and $g^m(v) = \beta$. We let $I_{2k-1}^{(m+1)}$ (resp. $I_{2k}^{(m+1)}$) be the closed interval with end-points s and t (resp. with end-points u and v). Then $I_{2k-1}^{(m+1)} \cup I_{2k}^{(m+1)} \subset I_k^{(m)}$ and $g^{m+1}(I_j^{(m+1)}) \supset [\alpha,\beta]$ for $j = 2k-1, 2k$. In this way we can thus construct intervals $I_k^{(m)}$, $1 \leq k \leq 2^m$, $m \geq 1$, having the required properties. Let

$$S = \{ \{\varepsilon_n\}_{n\geq 1} : \varepsilon_n \in \{0,1\} \text{ for each } n \geq 1 \};$$

we will use the intervals $I_k^{(m)}$ to define an injective mapping $\gamma : S \to \overline{W}$. This will show that $\overline{W}$ is uncountable, because S is uncountable. Note that each of the intervals $I_k^{(m)}$ contains an element of W. Fix $z = \{\varepsilon_n\}_{n\geq 1} \in S$, and for $m \geq 1$ let $\mu_m = 1 + \sum_{j=1}^{m} 2^{m-j}\varepsilon_j$; then $1 \leq \mu_m \leq 2^m$, and it is easy to see that $I_{\mu_{m+1}}^{(m+1)} \subset I_{\mu_m}^{(m)}$ for each $m \geq 1$. Put $D(z) = \bigcap_{m\geq 1} I_{\mu_m}^{(m)}$; then, since $I_{\mu_m}^{(m)} \cap W \neq \varnothing$ for each $m \geq 1$, we have $D(z) \cap \overline{W} \neq \varnothing$, and hence we can define a mapping

$\gamma : S \to \overline{W}$ so that $\gamma(z) \in D(z) \cap \overline{W}$ for each $z \in S$. Now let $z = \{\varepsilon_n\}_{n\geq 1}$, $z' = \{\varepsilon'_n\}_{n\geq 1} \in S$ with $z \neq z'$, and let $m = \min\{ n \geq 1 : \varepsilon_n \neq \varepsilon'_n \}$. Then

$$\mu_m = 1 + \sum_{j=1}^{m} 2^{m-j}\varepsilon_j \neq \mu'_m = 1 + \sum_{j=1}^{m} 2^{m-j}\varepsilon'_j ,$$

and $\gamma(z) \in I^{(m)}_{\mu_m}$, $\gamma(z') \in I^{(m)}_{\mu'_m}$; thus $\gamma(z) \neq \gamma(z')$. Therefore γ is injective, and so $\overline{W}$ is uncountable. □

REFERENCES

Block, L. (1977): *Mappings of the interval with finitely many periodic points have zero entropy.* Proc. of the A.M.S., 67, 357-360 .

Block, L. (1979): *Simple periodic orbits of mappings of the interval.* Trans. of the A.M.S., 254, 391-398 .

Block, L., J. Guckenheimer, M. Misiurewicz and L.-S. Young (1980): *Periodic points and topological entropy of one dimensional maps.* Global Theory of Dynamical Systems. Proceedings 1979. Z. Nitecki and C. Robinson, eds., Springer Lecture Notes in Math., 819, 18-34 .

Carathéodory, C. (1918): *Vorlesungen Uber reelle Funktionen.* Reprinted by Chelsea Publishing Company (1968), New York .

Collet, P. and J.-P. Eckmann (1980): *Iterated maps on the interval as dynamical systems.* Progress in Physics, Vol. 1 , Birkhäuser, Boston .

Collet, P., J.-P. Eckmann and O. Lanford (1980): *Universal properties of maps on an interval.* Comm. Math. Phys., 76, 211-254 .

Feigenbaum, M. (1978) and (1979): *Quantative universality for a class of nonlinear transformations.* J. Stat. Phy., 19, 25-52; 21, 669-706 .

Guckenheimer, J. (1979): *Sensitive dependence to initial conditions for one-dimensional maps.* Comm. Math. Phys., 70, 133-160 .

Guckenheimer, J., G. Oster and A. Ipaktchi (1977): *Dynamics of density dependent population models.* J. Math. Bio., 4, 101-147 .

Gumowski, I. and C. Mira (1980): *Recurrences and Discrete Dynamical Systems.* Springer Lecture Notes in Mathematics, Vol. 809 .

Hofbauer, F. (1981): *The structure of piecewise monotonic transformations.* Ergod. Theory and Dynam. Sys., 1, 159-178 .

Hofbauer, F. (1986): *Piecewise invertible dynamical systems.* Prob. Theory Rel. Fields, 72, 359-386 .

Jonker, L. and D. Rand (1981): *Bifurcations in one dimension, I: The nonwandering set,* and *II: A versal model for bifurcations.* Invent. math., 62, 347-365, and 63, 1-16 .

Lorenz, E. (1963): *Deterministic nonperiodic flow.* J. Atmos. Sci., 20, 130-141 .

May, R. (1976): *Simple mathematical models with very complicated dynamics.* Nature, 261, 459-467 .

May, R. and G. Oster (1976): *Bifurcations and dynamic complexity in simple biological models.* Amer. Nat., 110, 573-599 .

Milnor, J. and W. Thurston (1977): *On iterated maps of the interval. I. The kneading matrix,* and *II. Periodic points.* Preprint, Princeton University .

Misiurewicz, M. (1980): *Invariant measures for continuous transformations of* $[0,1]$ *with zero topological entropy.* Ergodic Theory Proceedings, 1978. M. Denker and K. Jacobs, eds., Springer Lecture Notes in Math., 729, 144-152 .

Misiurewicz, M. (1981): *Absolutely continuous measures for certain maps of an interval.* Publ. Math. I.H.E.S., 53, 17-51 .

Misiurewicz, M. and W. Szlenk (1980): *Entropy of piecewise monotone maps.* Studia Math., 67, 45-63 .

Nitecki, Z. (1982): *Topological dynamics on the interval.* Ergodic Theory and Dynamical Systems II, Proc., Special Year Maryland 1979-80. A. Katok, ed., Progress in Math., Vol. 21 , Birkhäuser, Boston .

Oxtoby, J. (1971): *Measure and Category.* Graduate Texts in Math., Vol. 2 , Springer-Verlag, New York Heidelberg Berlin .

Parry, W. (1966): *Symbolic dynamics and transformations of the unit interval.* Trans. of the A.M.S., 122, 368-378 .

Preston, C. (1983): *Iterates of maps on an interval.* Springer Lecture Notes in Mathematics, Vol. 999 .

Rudin, W. (1964): *Principles of Mathematical Analysis.* McGraw-Hill, New York .

Šarkovskii, A. (1964): *Coexistence of cycles of a continuous map of the line into itself.* Ukr. Mat. Z., 16, 61-71 .

Štefan, P. (1977): *A theorem of Sarkovskii on the coexistence of periodic orbits of continuous endomorphisms of the real line.* Comm. Math. Phys., 54, 237-248 .

van Strien, S. (1981): *On the bifurcations creating horseshoes.* Dynamical Systems and Turbulence, Proceedings Warwick 1980. D. Rand and L.-S. Young, eds., Springer Lecture Notes in Math., 898, 316-351 .

Walters, P. (1982): *An Introduction to Ergodic Theory.* Grad. Texts in Math., Vol. 79 , Springer-Verlag, New York Heidelberg Berlin .

Willms, J. (1988): *Asymptotic behaviour of iterated piecewise monotone maps.* Ergod. Theory and Dynam. Sys., 8, 111-131 .

INDEX

LECTURE NOTES IN MATHEMATICS

Edited by A. Dold and B. Eckmann

Some general remarks on the publication of monographs and seminars

In what follows all references to monographs, are applicable also to multiauthorship volumes such as seminar notes.

§1. Lecture Notes aim to report new developments - quickly, informally, and at a high level. Monograph manuscripts should be reasonably self-contained and rounded off. Thus they may, and often will, present not only results of the author but also related work by other people. Furthermore, the manuscripts should provide sufficient motivation, examples and applications. This clearly distinguishes Lecture Notes manuscripts from journal articles which normally are very concise. Articles intended for a journal but too long to be accepted by most journals, usually do not have this "lecture notes" character. For similar reasons it is unusual for Ph.D. theses to be accepted for the Lecture Notes series.

Experience has shown that English language manuscripts achieve a much wider distribution.

§2. Manuscripts or plans for Lecture Notes volumes should be submitted either to one of the series editors or to Springer-Verlag, Heidelberg. These proposals are then refereed. A final decision concerning publication can only be made on the basis of the complete manuscripts, but a preliminary decision can usually be based on partial information: a fairly detailed outline describing the planned contents of each chapter, and an indication of the estimated length, a bibliography, and one or two sample chapters - or a first draft of the manuscript. The editors will try to make the preliminary decision as definite as they can on the basis of the available information.

§3. Lecture Notes are printed by photo-offset from typed copy delivered in camera-ready form by the authors. Springer-Verlag provides technical instructions for the preparation of manuscripts, and will also, on request, supply special staionery on which the prescribed typing area is outlined. Careful preparation of the manuscripts will help keep production time short and ensure satisfactory appearance of the finished book. Running titles are not required; if however they are considered necessary, they should be uniform in appearance. We generally advise authors not to start having their final manuscripts specially tpyed beforehand. For professionally typed manuscripts, prepared on the special stationery according to our instructions, Springer-Verlag will, if necessary, contribute towards the typing costs at a fixed rate.

The actual production of a Lecture Notes volume takes 6-8 weeks.

.../...

§4. Final manuscripts should contain at least 100 pages of mathematical text and should include

- a table of contents
- an informative introduction, perhaps with some historical remarks. It should be accessible to a reader not particularly familiar with the topic treated.
- a subject index; this is almost always genuinely helpful for the reader.

§5. Authors receive a total of 50 free copies of their volume, but no royalties. They are entitled to purchase further copies of their book for their personal use at a discount of 33.3 %, other Springer mathematics books at a discount of 20 % directly from Springer-Verlag.

Commitment to publish is made by letter of intent rather than by signing a formal contract. Springer-Verlag secures the copyright for each volume.

Vol. 1173: H. Delfs, M. Knebusch, Locally Semialgebraic Spaces. XVI, 329 pages. 1985.

Vol. 1174: Categories in Continuum Physics, Buffalo 1982. Seminar. Edited by F.W. Lawvere and S.H. Schanuel. V, 126 pages. 1986.

Vol. 1175: K. Mathiak, Valuations of Skew Fields and Projective Hjelmslev Spaces. VII, 116 pages. 1986.

Vol. 1176: R.R. Bruner, J.P. May, J.E. McClure, M. Steinberger, H_∞ Ring Spectra and their Applications. VII, 388 pages. 1986.

Vol. 1177: Representation Theory I. Finite Dimensional Algebras. Proceedings, 1984. Edited by V. Dlab, P. Gabriel and G. Michler. XV, 340 pages. 1986.

Vol. 1178: Representation Theory II. Groups and Orders. Proceedings, 1984. Edited by V. Dlab, P. Gabriel and G. Michler. XV, 370 pages. 1986.

Vol. 1179: Shi J.-Y. The Kazhdan-Lusztig Cells in Certain Affine Weyl Groups. X, 307 pages. 1986.

Vol. 1180: R. Carmona, H. Kesten, J.B. Walsh, École d'Été de Probabilités de Saint-Flour XIV – 1984. Édité par P.L. Hennequin. X, 438 pages. 1986.

Vol. 1181: Buildings and the Geometry of Diagrams, Como 1984. Seminar. Edited by L. Rosati. VII, 277 pages. 1986.

Vol. 1182: S. Shelah, Around Classification Theory of Models. VII, 279 pages. 1986.

Vol. 1183: Algebra, Algebraic Topology and their Interactions. Proceedings, 1983. Edited by J.-E. Roos. XI, 396 pages. 1986.

Vol. 1184: W. Arendt, A. Grabosch, G. Greiner, U. Groh, H.P. Lotz, U. Moustakas, R. Nagel, F. Neubrander, U. Schlotterbeck, One-parameter Semigroups of Positive Operators. Edited by R. Nagel. X, 460 pages. 1986.

Vol. 1185: Group Theory, Beijing 1984. Proceedings. Edited by Tuan H.F. V, 403 pages. 1986.

Vol. 1186: Lyapunov Exponents. Proceedings, 1984. Edited by L. Arnold and V. Wihstutz. VI, 374 pages. 1986.

Vol. 1187: Y. Diers, Categories of Boolean Sheaves of Simple Algebras. VI, 168 pages. 1986.

Vol. 1188: Fonctions de Plusieurs Variables Complexes V. Séminaire, 1979–85. Edité par François Norguet. VI, 306 pages. 1986.

Vol. 1189: J. Lukeš, J. Malý, L. Zajíček, Fine Topology Methods in Real Analysis and Potential Theory. X, 472 pages. 1986.

Vol. 1190: Optimization and Related Fields. Proceedings, 1984. Edited by R. Conti, E. De Giorgi and F. Giannessi. VIII, 419 pages. 1986.

Vol. 1191: A.R. Its, V.Yu. Novokshenov, The Isomonodromic Deformation Method in the Theory of Painlevé Equations. IV, 313 pages. 1986.

Vol. 1192: Equadiff 6. Proceedings, 1985. Edited by J. Vosmansky and M. Zlámal. XXIII, 404 pages. 1986.

Vol. 1193: Geometrical and Statistical Aspects of Probability in Banach Spaces. Proceedings, 1985. Edited by X. Fernique, B. Heinkel, M.B. Marcus and P.A. Meyer. IV, 128 pages. 1986.

Vol. 1194: Complex Analysis and Algebraic Geometry. Proceedings, 1985. Edited by H. Grauert. VI, 235 pages. 1986.

Vol.1195: J.M. Barbosa, A.G. Colares, Minimal Surfaces in $\mathbb{R}^3$. X, 124 pages. 1986.

Vol. 1196: E. Casas-Alvero, S. Xambó-Descamps, The Enumerative Theory of Conics after Halphen. IX, 130 pages. 1986.

Vol. 1197: Ring Theory. Proceedings, 1985. Edited by F.M.J. van Oystaeyen. V, 231 pages. 1986.

Vol. 1198: Séminaire d'Analyse, P. Lelong – P. Dolbeault – H. Skoda. Seminar 1983/84. X, 260 pages. 1986.

Vol. 1199: Analytic Theory of Continued Fractions II. Proceedings, 1985. Edited by W.J. Thron. VI, 299 pages. 1986.

Vol. 1200: V.D. Milman, G. Schechtman, Asymptotic Theory of Finite Dimensional Normed Spaces. With an Appendix by M. Gromov. VIII, 156 pages. 1986.

Vol. 1201: Curvature and Topology of Riemannian Manifolds. Proceedings, 1985. Edited by K. Shiohama, T. Sakai and T. Sunada. VII, 336 pages. 1986.

Vol. 1202: A. Dür, Möbius Functions, Incidence Algebras and Power Series Representations. XI, 134 pages. 1986.

Vol. 1203: Stochastic Processes and Their Applications. Proceedings, 1985. Edited by K. Itô and T. Hida. VI, 222 pages. 1986.

Vol. 1204: Séminaire de Probabilités XX, 1984/85. Proceedings. Edité par J. Azéma et M. Yor. V, 639 pages. 1986.

Vol. 1205: B.Z. Moroz, Analytic Arithmetic in Algebraic Number Fields. VII, 177 pages. 1986.

Vol. 1206: Probability and Analysis, Varenna (Como) 1985. Seminar. Edited by G. Letta and M. Pratelli. VIII, 280 pages. 1986.

Vol. 1207: P.H. Bérard, Spectral Geometry: Direct and Inverse Problems. With an Appendix by G. Besson. XIII, 272 pages. 1986.

Vol. 1208: S. Kaijser, J.W. Pelletier, Interpolation Functors and Duality. IV, 167 pages. 1986.

Vol. 1209: Differential Geometry, Peñíscola 1985. Proceedings. Edited by A.M. Naveira, A. Ferrández and F. Mascaró. VIII, 306 pages. 1986.

Vol. 1210: Probability Measures on Groups VIII. Proceedings, 1985. Edited by H. Heyer. X, 386 pages. 1986.

Vol. 1211: M.B. Sevryuk, Reversible Systems. V, 319 pages. 1986.

Vol. 1212: Stochastic Spatial Processes. Proceedings, 1984. Edited by P. Tautu. VIII, 311 pages. 1986.

Vol. 1213: L.G. Lewis, Jr., J.P. May, M. Steinberger, Equivariant Stable Homotopy Theory. IX, 538 pages. 1986.

Vol. 1214: Global Analysis – Studies and Applications II. Edited by Yu.G. Borisovich and Yu.E. Gliklikh. V, 275 pages. 1986.

Vol. 1215: Lectures in Probability and Statistics. Edited by G. del Pino and R. Rebolledo. V, 491 pages. 1986.

Vol. 1216: J. Kogan, Bifurcation of Extremals in Optimal Control. VIII, 106 pages. 1986.

Vol. 1217: Transformation Groups. Proceedings, 1985. Edited by S. Jackowski and K. Pawalowski. X, 396 pages. 1986.

Vol. 1218: Schrödinger Operators, Aarhus 1985. Seminar. Edited by E. Balslev. V, 222 pages. 1986.

Vol. 1219: R. Weissauer, Stabile Modulformen und Eisensteinreihen. III, 147 Seiten. 1986.

Vol. 1220: Séminaire d'Algèbre Paul Dubreil et Marie-Paule Malliavin. Proceedings, 1985. Edité par M.-P. Malliavin. IV, 200 pages. 1986.

Vol. 1221: Probability and Banach Spaces. Proceedings, 1985. Edited by J. Bastero and M. San Miguel. XI, 222 pages. 1986.

Vol. 1222: A. Katok, J.-M. Strelcyn, with the collaboration of F. Ledrappier and F. Przytycki, Invariant Manifolds, Entropy and Billiards; Smooth Maps with Singularities. VIII, 283 pages. 1986.

Vol. 1223: Differential Equations in Banach Spaces. Proceedings, 1985. Edited by A. Favini and E. Obrecht. VIII, 299 pages. 1986.

Vol. 1224: Nonlinear Diffusion Problems, Montecatini Terme 1985. Seminar. Edited by A. Fasano and M. Primicerio. VIII, 188 pages. 1986.

Vol. 1225: Inverse Problems, Montecatini Terme 1986. Seminar. Edited by G. Talenti. VIII, 204 pages. 1986.

Vol. 1226: A. Buium, Differential Function Fields and Moduli of Algebraic Varieties. IX, 146 pages. 1986.

Vol. 1227: H. Helson, The Spectral Theorem. VI, 104 pages. 1986.

Vol. 1228: Multigrid Methods II. Proceedings, 1985. Edited by W. Hackbusch and U. Trottenberg. VI, 336 pages. 1986.

Vol. 1229: O. Bratteli, Derivations, Dissipations and Group Actions on C*-algebras. IV, 277 pages. 1986.

Vol. 1230: Numerical Analysis. Proceedings, 1984. Edited by J.-P. Hennart. X, 234 pages. 1986.

Vol. 1231: E.-U. Gekeler, Drinfeld Modular Curves. XIV, 107 pages. 1986.